Jung's Personality Theory Quantified

Douglass J. Wilde

Jung's Personality Theory Quantified

Springer

Prof. Douglass J. Wilde
ME Design Group
Stanford University
Building 550, Room 114
Escondido Mall 416, Stanford
CA 94305-4021, USA
e-mail: wilde@stanford.edu

ISBN 978-0-85729-099-1 e-ISBN 978-0-85729-100-4

DOI 10.1007/978-0-85729-100-4

Springer London Dordrecht Heidelberg New York

British Library Cataloguing in Publication Data
A catalogue record for this book is available from the British Library

Library of Congress Control Number: 2010938118

Cover design: eStudio Calamar S.L.

Printed on acid-free paper

Springer is part of Springer Science+Business Media (www.springer.com)

To my wife Jane, still a good date!

Preface

This work intends to fill an urgent need for managers, consultants, counselors, teachers, psychoanalysts, human resource professionals and others who use personality questionnaires, especially the Myers-Briggs Type Indicator *(MBTI)*. The need arises from the recent exposure of approximations and logical flaws in the traditional "type dynamics" procedure for determining the creative problem-solving elements of C. G. Jung's famous personality theory known as "function-attitudes", relabeled here as "cognitive modes". This book details a theory for finding these cognitive modes that is free from these approximations and logical errors. The new theory is fully QUANTitative, using the *MBTI* response counts long avoided by the entirely QUALitative approach of type dynamics.

The new quantitative theory is based on three fundamental postulates that follow from Jung's writings. The first two postulates, concerning respectively extraversion and the psychological functions for perception and judgment, are well known and widely used. The third, that the extraversion energy associated with the perception function is different from that of the judgment function, is here for the first time made explicit and quantitative. This contrasts with Jung's expression of it, which being implicit was entirely qualitative.

Quantification of these two different and independent extraversion energies is achieved by recognizing the mathematical coupling between Jung's *E-I* attitude and Myers' *P-J* attitude. "Coupling" refers to the strong interaction between the *E-I* and *P-J* attitudes, contrasting with the independence required of the two attitude variables that Jung envisioned but did not formulate explicitly. This undesirable interaction is removed by generating two new "decoupled" attitudes completely independent of each other.

The scores for these new attitudes—extraverted vs. introverted perception on the one hand and extraverted vs. introverted judgment on the other—are calculated simply as an average sum and an average difference of the original *E-I* and *P-J* attitude scores. Alternatively, the average sum may be subtracted from the larger coupled attitude score to "unpack" it into two decoupled attitudes, one in each domain. Although this simple linear transformation may seem mysterious to many *MBTI* users, it is well known to systems analysts.

Decoupling is worked out in the fourth chapter, after the introductory Chap. 1, a review of Jung's present qualitative personality theory in Chap. 2, and the quantitative preliminaries of Chap. 3. Chapter 3 allows comparison of scores from various personality instruments by expressing them as percentages of their ranges. This normalization turns out to simplify not only the attitude decoupling of Chap. 4, but also the combination in Chap. 5 of attitude and function scores to generate cognitive mode scores. Decoupling reduces the dimensionality of *MBTI* from four, for which no graphical treatment is conceivable, to two dimensions for perception and two others for judgment. This allows each domain to be described by ordinary two-dimensional graphs permitting graphical analysis of personality, an advantage not possessed by any other personality instrument.

Surprisingly perhaps, the proof strategies of Chap. 4 carry over into Chap. 5 for matching *MBTI* score combinations with those for the modes. This correspondence is needed to construct a rigorous mapping between the *MBTI* and the mode scores. Just as the Greek philosopher Plato envisioned the "ideal" citizen of his *Republic* as blending certain "ideal" qualities of character, so Chaps. 4 and 5 see the "ideal" extraverted sensor (for example) as combining "ideal" (maximum) values of both the extraverted perception attitude and the sensing function. This approach allows the generation of non-ideal scores by very simple interpolation formulas.

The construction reveals personality potentials hidden from the limited type dynamics approach. It is a straightforward mathematical principle that the two variables, attitude and function, from which each cognitive mode is formed, must generate scores for two modes, not just one, in both the perception and the judgment domains. In other words, each of the four pairs of cognitive modes can contribute a significant mode, as many as four in all. Usually some of these scores are numerically insignificant, but a considerable number of students in the author's classes have had three or even four significant modes, of which only two can be detected by type dynamics. Such extra "hidden" modes expand a person's creative potential and widen the range of team roles or other activities to which he can be assigned. This is an important principle for situations described in the author's *Teamology: The Construction and Organization of Effective Teams* (2009, Springer London). Thus the *MBTI*, properly analyzed quantitatively, contains up to twice the personality information extractable by type dynamics. Mathematical psychologists may well recognize other psychological theories combining two variables into one while overlooking the inevitable second.

Chapter 6 examines the impact of these developments not only on Jung's dominant and auxiliary modes, but also on the newly identified "subsidiary" modes overlooked by type dynamics. Counselors and teachers in particular will find this chapter interesting because of the creative potential, previously lost, to be discovered in their students and clients. Quantitative considerations may alter the analyst's opinion of which modes are dominant and auxiliary, and there can be a "grey area" where dominance is ambiguous.

Type dynamics earns its own Chap. 7 examining its approximations and categorical reasoning. After saluting the pioneers of typology for their imagination and enterprise, the chapter explains the oversimplifications and perils of the categorical reasoning underlying type dynamics as criticized by experimental psychologist J. Reynierse. Four type dynamics assumptions are identified and then illustrated with numerical examples—and counterexamples. In what should be a relief to all *MB* personologists, the traditional confusion of the existing Myers-Briggs *P-J* attitude terminology is shown to be cleared up by that of the new decoupled attitudes. An important concept challenged is the type dynamics idea that people of the same Myers-Briggs types all have similar personalities. This notion is contradicted by five numerical examples of people with the same *ENTP* type but significantly different cognitive mode compositions.

Chapter 8 departs from the quantitative emphasis of the rest of the book in dealing with and defending Jungian analyst John Beebe's qualitative archetype theory for describing the unconscious, shadow portion of the human psyche. Being based entirely on clinical experience, the model cannot be verified experimentally in the manner applicable to the ego-based cognitive modes. Archetype theory can however guide not only psychoanalysts having other clinical information, but also lay consultants and managers dealing only with anecdotal incidents of interest to clients or teams. The position is taken that Beebe's theory can still be useful when it is freed from its original dependence on the errors and approximations of type dynamics.

The concluding Chap. 9 first reviews the extensions to Jung's earlier qualitative theory brought about by the book's quantitative, logical, and axiomatic approach. Then it suggests implications for other personality theories and questionnaires: the Singer-Loomis Inventory of Personality (SLIP), the *MBTI* Step II, and Five-Factor Analysis (FFA). It concludes with remarks about measuring the cognitive modes directly.

This book gives better methods for mapping the *MBTI* quantitatively to Jung's cognitive modes. Along the way it furthers Jung's original concepts while placing them on a solid axiomatic foundation not possessed by other personality theories. Jungian psychoanalysts may find this helpful in organizing complicated clinical information.

Bringing these quantitative findings to the millions of lay *MBTI* users worldwide will require further education of those already certified to administer the instrument. For this reason numerical exercises follow many of the chapters to make the book a source reference for briefer workbooks usable in enhanced certification programs. Backed by quantitative theory and new graphical methods, the pioneering qualitative typology work of Myers and Briggs is thus corrected and extended to yield deeper understanding of the vital topics of human personality, creativity and human relations.

Contents

Chapter 1
Introduction: Typology at a Crossroad

I took the (road) less travelled by,
And that has made all the difference.
–Robert Frost (1920)

The Myers–Briggs type indicator (MBTI), a famous questionnaire for measuring the variables of psychiatrist C. G. Jung's well-established personality theory, has been a cornerstone, at times controversial, of what has come to be known as "typology", a paradigm for studying and understanding human personality. Numbering in the hundreds of millions, its users may be shocked to learn that in 2009 existing methods for interpreting MBTI results in Jungian terms were challenged on both logical and statistical grounds. The validity of the MBTI itself was not questioned; it was rather the traditional "type dynamics" (TD) method for mapping the MBTI on to Jung's theory that was criticized. As this is written, passionate controversy over the matter threatens the unity of the community of not only the certified MBTI administrators and consultants, but also the millions employing it for career guidance, corporate job assignment, and psychological counseling.

This book brings to bear on the interpretation problem a novel approach that appears to resolve these unfortunate disputes. The approach actually strengthens the MBTI by bringing forth elements of Jungian theory that stand unintentionally outside the limited purview of TD. One can say without exaggeration that the new approach doubles the power of the MBTI to identify significant Jungian elements.

The main novelty of this approach is its use of the numerical data generated by the instrument. This contrasts with TD's dependence entirely on the MBTI's four qualitative type categories, with no exploitation of the numbers of responses to the questions. The traditional method of course loses information that is easily recovered by the new theory.

The novel QUANTitative theory also strengthens Jungian theory by putting it on an axiomatic basis from which important typological properties can be deduced logically, in the manner of Euclid's classical geometry, with no need for experimental verification. Experimental evidence can be regarded as testing the underlying axioms rather than the facts derived from them. It is hoped that dispassionate

D. J. Wilde, *Jung's Personality Theory Quantified*,
DOI: 10.1007/978-0-85729-100-4_1, © Springer-Verlag London Limited 2011

analysis will defuse the current controversy and bring the MBTI community together again on a firm scientific basis.

1.1 History and Preview

Sections 1.2 through 1.7 review the 90-year history of Jung's Personality Theory. Starting with Jung's original qualitative theory in Sect. 1.2 and the development of the MBTI by Katherine Briggs and her daughter Isabel Myers in Sect. 1.3, the chapter describes TD in Sect. 1.4 and the eight-function model in Sect. 1.5 to which TD has been applied by advanced users. After the author's "teamology" application in Sect. 1.6 comes a brief account of the sensational challenge to TD made by J. Reynierse in 2009.

Section 1.8 previews the many new developments described in this book. These include a breaking down of the original four-dimensional problem into two systems of only two variables each. This decomposition, achieved by decoupling the attitudes confusedly linked together in the current model, for the first time allows personalities to be described correctly on graphs. Attitude decoupling also simplifies conceptualization of the Jungian elements ultimately sought. New function-attitudes are found that were hidden from TD theory. J. Beebe's clinical archetype theory, which some think had been questioned along with TD, is rehabilitated by separating it from the TD rules. All this is applied to the teaching of personality theory through the construction of creative problem-solving teams. Finally, extensions to Jung's theory and implications for other personality theories are discussed. Intending to reunite the MBTI community through re-education courses, the book has exercises following many of the chapters.

1.2 Jung's Qualitative Personality Theory

In his seminal book *Personality Theory,* pioneering psychiatrist Carl Gustav Jung distinguished between "attitudes" and "functions" of consciousness. In doing so he postulated two attitude pairs and two function pairs. Only one attitude pair—*Extraversion* versus *Introversion*—was explicit in his work, however. As discussed in the next section, the second pair came later. The function pairs by the way were named *Sensing* versus *iNtuition* (the "perception" functions) and *Thinking* versus *Feeling* (the "judgment" functions).

Using terminology developed for these three pairs, Jung then defined eight "function-attitudes", each combining exactly one function with one attitude, "introverted thinking" for example. Writers occasionally drop the "attitude" part, which unfortunately can confuse the hyphenated entity with the totally different variable called "function". To avoid such category mistakes, this book will call them "cognitive modes" as suggested by Jungian analysts Singer and Loomis (1984).

1.3 Measurement with the MBTI

Intending to measure Jung's variables, his followers constructed a questionnaire known as the Grey-Wheelwrights (GW) instrument (Wheelwright et al. 1964). Reflecting the state of Jung's early theory, the GW still lacked one attitude pair. Concurrently and independently, Katherine Briggs had studied another pair of personality variables that eventually became the missing second attitude pair. She and her daughter Isabel Myers combined their new *Perception* versus *Judgment* variable with the GW to construct the MBTI instrument having four pairs of variables: *E-I*, *S-N*, and *T-F* from GW, together with the new attitude *P-J* (Myers 1975).

Since the names Briggs and Myers gave their new attitude pair had already been used to describe sets of functions, only the initial letters will be used here in order to reduce categorical confusion, a subject developed more fully in Sect. 7.3. This terminology problem will be eliminated in Chap. 4 by the introduction of two new "decoupled" attitudes replacing both *E-I* and *P-J*.

Many books have explored the pairs individually, guiding a profession of type consultants who advise clients about avoiding and dealing with type conflicts. "Type-watching" has become a game-like activity not without value in human relations.

Myers not only tested and developed the MBTI, she also wrote much to promote it as a popular and easy way to understand personalities and their interactions. Her lucid type descriptions and her book *Gifts Differing* written with her son Peter (1980) attracted several hundred million people to the MBTI and the related doctrine of learning to respect personality differences instead of rejecting them.

1.4 Type Dynamics

To generate her famous 4×4 type table of descriptions for the 16 "types"—combinations of the four type letter pairs—Myers developed rules using the letter categories but not the response counts to identify the associated dominant and auxiliary cognitive modes (Myers et al. 1989, 1998). This rule-based system, along with other rules for estimating the attitudes associated with the two non-preferred functions, has come to be known as "TD". TD is the vortex of the controversy described in Sect. 1.7.

1.5 The Eight Function(-Attitude) Model

With the estimation of cognitive modes made simple by the TD rules, many type consultants expanded their practices by basing them on cognitive modes as well as, or even instead of, the primitive MBTI variables (Thompson 1996, Haas,

McAlpine and Hartzler 2001, Haas and Hunziker 2006). This "eight-function model", as it has been called, has opened up many fascinating applications to interpersonal relations. Strictly speaking, it refers to function-*attitudes*, not functions, but there's no need to quibble about that here. To avoid categorical confusion however, the terminology "eight-*mode* model" will be employed in the sequel. Notice that the needed determination of the modes depends on the very TD rules challenged in Sect. 1.7.

1.6 Teamology

Meanwhile, the author was applying the eight-mode model to the problem of constructing student design teams. He found that the most effective teams were those whose members' cognitive mode preferences together encompassed all eight modes. Constructing such teams therefore required knowing the favored modes for every student, a formidable task for a class of thirty or more if the complicated rules of TD were employed. To save time he devised a spreadsheet based on the response difference counts overlooked by TD, reasoning that the quantitative data would contain at least as much typological information as the letters of the type code.

This "teamology" reasoning laid the foundation for the quantification strategy described in Chaps. 4 and 5 of this book. Its mode findings agreed with those of TD most of the time, and when they disagreed, further investigation confirmed the quantitative prediction enough that eventually it was used exclusively. Then after a few years it was realized that the quantitative spreadsheet, which automatically examines all four of Jung's mode pairs, was often detecting significant preferences for three or even all four modes, not just the two identified by TD. It turned out that these extra modes, called "subsidiary" in this book's Chap. 5, were indeed valid preferences significant enough to guide the assignment of students to duties on a team. Such assignments in fact improved team performances even further, leading to the publication of *Teamology* (Wilde 2009).

While *Teamology* was in press, the author realized that its main value may not have been so much the team success it described as it was the newly—and almost accidentally—discovered transformation of the quantitative data used to identify the cognitive mode patterns. This insight motivated the author to start bringing the transformation to the attention of the MBTI community because of its power to find creative potentials hidden from traditional TD analysis.

1.7 Type Dynamics Challenged

Coincidentally, or as Jung might say, "synchronistically", the month *Teamology* appeared officially also saw the publication of a controversial article entitled

"The Case Against Type Dynamics" (Reynierse 2009). Preceded by 2 statistical studies of over 700 MBTI evaluations (Reynierse and Harker 2008), it objected not only to TD's failure to fit the data, but also to its inconsistencies, logical errors, incompleteness, and dependence on rules rather than rigorous theory. It moreover challenged much experimental evidence intended to support TD as statistically confounding TD with its MBTI content. As an alternative he proposed an approach called "preference multidimensionality" in which personality characteristics are studied in terms of combinations of two or more MBTI variables. Chapter 7 describes, as calmly as possible, this sensational challenge to TD.

The author immediately realized that the teamology transformation does not suffer from the deficiencies Reynierse noted and, for the moment at least, may be the "only game in town" for identifying the eight modes needed for team construction and so many other applications. This epiphany motivated the writing of this book.

So many TD users were offended by the article that those more research-minded banded together to organize a conference defending TD. As this is written a year later, this "Examining the Evidence" conference did not attract enough papers to be viable, but the seven articles submitted are being tentatively slated to accompany a meeting of the International Association of Psychological Type (APTi) in 2011. With luck this book will be published by then. May its independence from TD along with its ability to identify the needed modes reunify the presently troubled MBTI community.

1.8 Preview of New Developments

The rest of the book is devoted to many new developments based on quantifying Jung's cognitive mode personality model. Chapter 2 reviews Jung's qualitative theory, placing it on an axiomatic basis to allow its properties to be deduced formally in the manner of Euclid's classical geometry, with minimal psychological interpretation. To two well-established postulates involving extraversion and the psychological functions, a third is added requiring independence of the attitudes, a condition to be known in Chap. 4 as "attitude decoupling", a concept developed after Chap. 3 introduces some preliminaries needed for the quantitative work to follow.

Decoupling reduces the problem from a four-dimensional one to two independent systems each having only two variables. These two-dimensional systems are then easily analyzed separately for visualization by simple graphical methods developed in Chaps. 4 and 5. The theory identifies not only the cognitive modes expected by TD, but also one or two "subsidiary" modes hidden from TD (Chap. 6).

After expressing respect for the daring and enterprising MBTI pioneers, Chap. 7 discusses the traditional TD rules, along with Reynierse's critique of them, and suggests how to fix them. Beebe's clinical archetypes, unintentionally placed somewhat under a cloud by Reynierse, are in Chap. 8 rehabilitated by cutting them free from dependence on the TD rules Chap. 9 applies all this to teamology. In Chap. 10 the book concludes by reviewing its extension of Jung's qualitative personality theory and calling attention to further implications of this work for other type description systems.

One hopes this endeavor will reunite the MBTI community. Perhaps this text will prove useful as a workbook for new certification and RE-certification courses incorporating its novel and revealing quantitative approach to Jung's personality theory.

Chapter 2
Jung's Qualitative Personality Theory

This above all; to thine own self be true.
And it must follow, as the night the day,
Thou canst not then be false to any man.
==Hamlet, William Shakespeare. 1604

2.1 Introduction

Jung's personality theory, originally a guide for psychoanalysts seeking patterns in clients' psyches, has until now been entirely qualitative and free of numbers. Its quantification to come must be based on and be consistent with this qualitative description. This chapter places this informal set of descriptions on an axiomatic foundation so that typological "principles" can be deduced logically and rigorously as in Euclid's classical geometry. To make this accessible to the non-mathematical reader, the presentation will be kept informal rather than being expressed in the dry theorems and corollaries of high school.

The qualitative theory will be based on three premises called "postulates" from which later principles can be derived. The first two, concerning extraversion and the psychological functions, are explicit and already widely accepted. The third "domain independence" postulate, expressed in Jung's writings only implicitly, will be made explicit here to guide its quantification in the next two chapters.

Only the first two postulates are needed for combining attitudes with functions to generate Jung's personality types, later known in a more general context as "function-attitudes" or "cognitive modes". Jung's typology also includes a second "auxiliary" mode, sometimes undifferentiated with respect to attitude.

The chapter concludes by discussing the Grey–Wheelwrights questionnaire developed by Jung's followers as an attempt at mode identification. Based on only the first two postulates, its coverage of the conscious ego was incomplete for lack of the second attitude pair provided later by Briggs and Myers. An example will illustrate this earlier incompleteness.

2.2 Three Axiomatic Postulates

In what follows, in citations Jung's book *Personality Theory* (1921, 1971) will be abbreviated as "*PT*", usually followed by a page number.

D. J. Wilde, *Jung's Personality Theory Quantified*,
DOI: 10.1007/978-0-85729-100-4_2, © Springer-Verlag London Limited 2011

2.2.1 Postulate 1: Extraversion (and Introversion)

To understand and explain their differences in approach to the same problems by his Vienna colleagues Freud and Adler, Jung formulated his famous theory of extraversion and introversion. These opposite variables are examples of "psychological attitudes" and are expressed symbolically as the *E–I* pair.

Postulate 1 *Extraversion* is the flow of psychic energy *outward* toward the exterior world—"an outward turning of libido" (*PT*, p. 427), whereas *introversion* draws psychic energy *towards* one's interior psyche—"an inward turning of libido" (*PT*, p. 452).

You can gain further understanding of these ideas, as well as your own preferences in this regard, by examining the following five questions from p. 10 of the *Teamology* questionnaire (Wilde 2009). If you are so inclined, feel free to circle zero, one or two alternatives for each of the five questions and do the math.

*EI*1	You are more:	(*e*) sociable	(*i*) reserved
*EI*2	You are more:	(*e*) expressive	(*i*) contained
*EI*3	You prefer:	(*e*) groups	(*i*) individuals
*EI*4	You learn better by:	(*e*) listening	(*i*) reading
*EI*5	You are more:	(*e*) talkative	(*i*) quiet

From the qualitative point of view of Postulate 1, you would regard yourself as an extravert if you chose more (*e*)s than (*i*)s, an introvert if you chose more (*i*)s than (*e*)s, and undecided if you chose the same number of each. The quantitative point of view to be developed later would append the positive difference between the numbers of the responses to the attitude identifier. Thus three (*e*)s and two (*i*)s would yield a report of *E*1 for this questionnaire (Example 1 for the author, to be continued in Chap. 4). The score can also be expressed as a percentage of the total, here 20% *E*. If you happen to have your *E–I* score from the *MBTI*, you can multiply it by 100%/30 to get the percentage score. This pair of attitudes is an example of what will be called a "variable (pair)".

2.2.2 Postulate 2: Psychological Functions

Jung soon realized that extraversion or introversion could change according to whether information was being collected or, having been obtained, was guiding a decision. In the first situation of "perception", as he called it, he distinguished between two opposite ways of information collection: *S*ensing versus *iN*tuition*S*-*N*. The *sensing* function *S* "include(s) all perceptions by means of the sense organs" (*PT*, p. 518), whereas the *intuition* function *N* "is perception by means of the unconscious" (ibid).

In the second "judgment" situation, he saw two opposite ways of decision-making that he called *T*hinking versus *F*eeling *T–F*. By the *thinking* function *T* Jung means "… intellectual cognition and the forming of logical conclusions", whereas "*feeling* (*F*) is a function of subjective valuation" (ibid).

To understand these "psychological functions", you can examine (or even answer) the following two sets of five questions from the teamology questionnaire. First, the perception function questions.

*SN*1	You prefer the:	(*s*) concrete	(*n*) abstract
*SN*2	You prefer:	(*s*) fact-finding	(*n*) speculating
*SN*3	You are more:	(*s*) practical	(*n*) conceptual
*SN*4	You are more:	(*s*) hands-on	(*n*) theoretical
*SN*5	You prefer the:	(*s*) traditional	(*n*) novel

To continue Example 1, the author's choice of all (*n*)s indicates an "intuitive" preference, scoring 100% *N.*

The judgment function questions follow:

*TF*1	You prefer:	(*t*) logic	(*f*) empathy
*TF*2	You are more:	(*t*) truthful	(*f*) tactful
*TF*3	You are more:	(*t*) questioning	(*f*) accommodating
*TF*4	You are more:	(*t*) skeptical	(*f*) tolerant
*TF*5	Judges should be:	(*t*) impartial	(*f*) merciful

In Example 1, three (*t*)s and two (*f*)s score as 20% *T*, indicating a "thinking" preference. Each pair of functions is called a "variable (pair)", as was the attitude pair of Postulate 1.

Postulate 2 (1) There exists an opposite pair of psychological functions representing two different ways of collecting information (perception).

(2) There exists another opposite pair of psychological functions representing two different ways of making a decision (judgment).

Jung gave names and letter symbols to these two pairs of functions, something he did NOT do to the corresponding two pairs of attitudes of Postulate 3.

2.2.3 Postulate 3: Perception and Judgment Domains

In *PT* the third postulate is implicit rather than explicit. Rather than being stated in so many words, it has to be inferred from the way Jung organizes, in *PT*'s Chap. X, pp. 330–407, his discussion of the type variables and how they interact.

Postulate 3 The extraversion or introversion attitude energy for information collection (perception) is independent of and usually different from that for decision-making (judgment).

As the preceding subsection noted, Jung did not give names and symbols to the two pairs of attitudes of Postulate 3. To correct this oversight, define the perception attitude pair as *extraverted* versus *introverted perception* (*E*p–*I*p) and the judgment attitude pair as *extraverted* versus *introverted judgment* (*E*j–*I*j). These new attitude pairs will facilitate development of the quantitative theory to come, especially in Chap. 4, where in Table 4.1 of Sect. 4.5 they will receive less formal titles.

The word "usually" in the postulate expresses that in rare cases one can actually have the same attitude energies in both situations, a topic discussed in Chaps. 4 and 7. Although such an equality of perception and judgment attitude energies, albeit rare, may surprise some type theorists, the author has known students and a successful corporate executive who comfortably possess such attitudes.

Since only the single *E–I* attitude of Postulate 1 has been defined so far, another attitude is needed for completeness. Section 2.4.1 will illustrate by example why the variables introduced so far are insufficient for describing a consciousness. The missing attitude, supplied by Briggs and Myers who symbolized it by *P–J*, will appear in Sect. 2.4.2 to complete the attitude requirement.

Postulate 3 states that the two new attitude pairs *E*p–*I*p and *E*j–*I*j must be independent of each other in the sense that changing one should not affect the other. Since the single attitude *E–I* of Postulate 1 certainly affects the new attitudes, both for perception and for judgment, it must be interacting in some way with the attitude *P–J* not yet added to the system. In Chap. 4, *E–I* and *P–J* will be "decoupled" by systems analysis to produce the independent attitude pairs *E*p–*I*p and *E*j–*I*j. Remarkably, it will turn out that the *E–I* score is actually the net algebraic sum of the two decoupled extraversion/introversion attitude scores, as yet unknown, for *E*p–*I*p and *E*j–*I*j.

2.3 Types and Cognitive Modes (Function-Attitudes)

Jung then combined attitudes with functions to obtain his personality *types*. Combination was slightly restricted; the four perceptive types for instance came only from the two perceptive attitudes and functions, Similarly, the four judgment types came only from judgment attitudes and functions, yielding the eight Jungian types displayed in Table 2.1. The descriptive keywords (Wilde 2009, p. 13) are intended to humanize this perhaps overly technical exposition.

John Beebe continually emphasizes that Jung was referring to types of *consciousness*, not types of *people*. The distinction is that although an individual favoring one of the consciousness types is popularly regarded as that type of *person*, she will doubtless experience many other types of *consciousness* during her life. To avoid confusing the two interpretations, *consciousness* types will in

Table 2.1 Jungian types (cognitive modes) with *teamology* keywords

Perception modes			Judgment modes	
Extraverted sensing Se **Experiment**	*Extraverted iNtuition* Ne **Ideation**	–	*Extraverted thinking* Te **Organization**	*Extraverted feeling* Fe **Community**
Introverted sensing Si **Knowledge**	*Introverted iNtuition* Ni **Imagination**	–	*Introverted thinking* Ti **Analysis**	*Introverted feeling* Fi **Evaluation**

what follows be called "(cognitive) modes", whereas *people* types usually will be identified by an adjective such as "Jungian" or "Myers–Briggs" (*MB*) indicating an particular context and type definition. As noted in Sect. 1.5, of Chap. 1 "mode" here is synonymous with the term "function-attitude" favored by many of the eight-function(-attitude) school of typology.

Recognizing this multiplicity of cognitive modes, Jung distinguished between a "principal", later called "dominant", mode and a less differentiated or even undifferentiated "subsidiary" or "auxiliary" mode, one being perceptive and the other judgmental. Thus Jung described his own type as "introverted intuition with thinking", apparently regarding his auxiliary thinking function as being as often extraverted as introverted. This is as far as he would go using only the sort of information available to a Jungian psychoanalyst not using some sort of questionnaire. With this as a model, a "Jungian type" will be henceforth regarded as having a dominant function in a specific attitude, together with an auxiliary function whose attitude may or may not be differentiated.

As discussed in Sect. 7.3.3, of Chap. 7 one of the Type Dynamics rules always determines the auxiliary attitude so that a *TD* type has exactly two modes. The quantitative theory developed in Chap. 5 will generate positive scores for as many as four modes. Even though some of these may not be significant, quantification can represent an expansion of the power of the questionnaire to identify the correct cognitive modes, often more than two.

2.4 Questionnaires

2.4.1 The Jungian Type Survey (Grey–Wheelwrights)

Jung's qualitative ideas so far go a long way toward identifying a person's dominant mode after observing only the three pairs *E–I*, *S–N* and *T–F* qualitatively. As an example, take Jung himself. His quiet, scholarly demeanor would read as *I*ntroversion and *T*hinking, while his powerful imagination would point to *iN*tuition as the dominant function. Putting them together would indicate introverted intuition (Imagination) with *T*hinking auxiliary, the associated perception attitude being as yet unknown. No further information would appear to be needed for elementary psychoanalytic purposes.

Jung's followers soon sought to augment such clinical estimates with information obtained from a forced-choice questionnaire known as the Grey–Wheelwrights (*GW*) instrument (Wheelwright, Wheelwright and Buehler 1964). However, as foreshadowed in Sect. 2.2.3, the *GW* lacked the second attitude needed for full precision, but the remaining three variable pairs *E–I*, *S–N* and *T–F* could at least give a good idea about the dominant and auxiliary modes in much less time than the consulting room sessions that otherwise would have been needed.

Example 1's scores—20% *E*, 100% *N*, 20% *T*—would have been interpreted by the *GW* as indicating *E*xtraverted i*N*tuition with auxiliary *T*hinking of unknown attitude. Choosing *N* as dominant is based of course on its score (100%), much higher than that for the other function *T* (20%). This bit of mildly quantitative comparative reasoning, foreshadowing the major findings of this book, is completely absent from the analytic *TD* rationale of the famous *MBTI* questionnaire that followed the *GW* and eventually replaced it. Moreover, as would most Jungians of the day (Myers and Myers 1980, pp. 20–21), the Wheelwrights themselves would have assumed the auxiliary attitude to be the same as that of the dominant, i. e, extraverted (*E*j) for Jung and introverted (*I*j) for Example 1. Thus the *GW* would have seemed entirely sufficient to determine the dominant and auxiliary modes, attitudes and all. But this was not to be.

2.4.2 The Myers–Briggs Type Indicator (MBTI)

Briggs and Myers combined the missing second attitude *P–J* (Sects. 2.2.3 and 2.4.3) with *GW* style questions to develop the *MBTI*, on which Briggs' initial research slightly preceded even Jung's (ibid). Surprisingly perhaps, the new attitude doesn't by itself identify the auxiliary attitude as one might expect. This is because the traditional *MBTI* way of identifying the dominant and auxiliary modes involves only the four letter categories, not the response count data. For this reason the rules of what came to be known as Type Dynamics had to be introduced to produce definite conclusions.

The only *TD* rule relevant to the present discussion is "attitude balance", which states that the dominant and auxiliary attitudes must be opposite. In the examples this characterizes the auxiliary thinking modes as extraverted for Jung and introverted for Example 1. Chapter 4 will show how to resolve this by finding the *E*j–*I*j score as a combination of the *P–J* and *E–I* quantitative scores. Acknowledging the exceptions noted by the Jungians, even Myers and Myers point out (ibid) that attitude balance is not a hard and fast universal rule anyway. For M and M it seems to have been more a suggestion for "type development", a topic beyond the scope of this book. Sections 7.3.3 and 7.3.4 of Chap. 7 will further examine the logic of the rule and its alleged provenance from Jung himself, here respectfully disputed.

2.4.3 Briggs' P–J Attitude

So far the exposition has concerned only what the new and undefined *P–J* attitude pair *does* rather than what it *is*. Recall that to avoid the categorical error of naming these attitudes after the already named "perception" and "judgment" situations, they will be designated here only by the letters "*P*" and "*J*", pronounced "pea" and "jay" respectively whenever they need to be referenced verbally (Sect. 7.3 of Chap. 7).

The *P* attitude, here given the more descriptive keyword "flexibility", involves, in the words of *M and M*, pp. 71–72: spontaneity, open-mindedness, understanding, tolerance, curiosity, zest for experience and adaptability. Its opposite attitude *J*, with keyword "structure", involves (ibid, pp. 70–71): system, order, planning, sustained effort, decisiveness, authority, opinion and routine.

To understand this second pair of "psychological attitudes", you may wish to examine (or even answer) the following five questions from the teamology questionnaire.

*PJ*1	You are more:	(*p*) casual	(*j*) systematic
*PJ*2	You prefer activities to be:	(*p*) open-ended	(*j*) planned
*PJ*3	You work better:	(*p*) with pressure	(*j*) without pressure
*PJ*4	You prefer:	(*p*) variety	(*j*) routine
*PJ*5	You are more:	(*p*) improvisational	(*j*) methodical

With some embarrassment, the author must call attention to item *PJ*3, which was inadvertently reversed in the Teamology Questionnaire on p. 20 of (Wilde 2009). Teamologists, take note—this error can throw the final modes scores 10% off!

To continue Example 1, four (*p*)s and one (*j*) score as 60% *P*, indicating a "flexible" preference. You can decide if this flexibility suggests an auxiliary *extraverted* judgment attitude as it would to the Wheelwrights, or an auxiliary *introverted* judgment attitude in accord with the *TD* attitude balance rule. The corresponding auxiliary judgment modes are respectively *T*e Organization and *T*i Analysis.

Later experience (Myers et al. 1998) has established the correspondences shown in Table 2.2 between the cognitive modes and the personality variables. Notice that *P* occurs just as often in (introverted) judgment modes as in (extraverted) perception modes, and *J* occurs as often in (introverted) perception modes as in (extraverted) judgment modes. This would produce total confusion if *P*ea and *J*ay carried the names originally, and innocently, awarded them by Briggs and Myers. Section 4.3.4 of Chap. 4 will develop a clearer interpretation of *P–J* as distributing extraversion energy between the two domains.

Table 2.2 Cognitive modes with functions, attitudes and *teamology* keywords Briggs attitudes ***P–J*** in ***boldface***

Perception modes			Judgment modes	
Extraverted sensing Se: *E, S,* ***P*** Experiment	*Extraverted iNtuition* Ne: *E, N,* ***P*** Ideation	–	*Extraverted thinking* Te: *E, T,* ***J*** Organization	*Extraverted feeling* Fe: *E, F,* ***J*** Community
Introverted sensing Si: *I, S,* ***J*** Knowledge	*Introverted iNtuition* Ni; *I, N,* ***J*** Imagination	–	*Introverted thinking* Ti: *I, T,* ***P*** Analysis	*Introverted feeling* Fi: *I, F,* ***P*** Evaluation

Table 2.3 *MBTI* types and assumed cognitive modes

ISTJ Si, *Te*	*ISFJ* Si, *Fe*	*INFJ* Ni, *Fe*	*INTJ* Ni, *Te*
ISTP Ti, *Se*	*ISFP* Fi, *Se*	*INFP* Fi, *Ne*	*INTP* Ti, *Ne*
ESTP Se, *Ti*	*ESFP* Se, *Fi*	*ENFP* Ne, *Fi*	*ENTP* Ne, *Ti*
ESTJ Te, Si	*ESFJ Fe,* Si	*ENFJ Fe,* Ni	*ENTJ Te,* Ni

2.4.4 The Type Table

Although the new attitude *P–J* ultimately will bring much to the quantitative analysis of personality, Myers' really brilliant contribution to personology was her inspiration to write simple type descriptions using both dominant and auxiliary modes associated with the $2 \times 2 \times 2 \times 2 = 16$ four-letter types, henceforth known as the *MB* types. Table 2.3 shows these types together with the dominant and auxiliary cognitive modes associated with them according to the *TD* attitude balance rule.

Each type description had a few words describing the two modes, mostly about the dominant mode. For example, the *ENTP* description corresponding to Example 1's *ENTP* type reads (Myers et al. 1989, p. 21):

> *ENTP*: Quick, ingenious, good at many things. Stimulating company, alert and outspoken. May argue for fun on either side of a question. Resourceful in solving new and challenging problems. But may neglect routine assignments. Able to form one new interest after another. Skillful at finding logical reasons for what they want.

Here the dominant *E*xtraverted *iN*tuition *N*emode contributes the words in **boldface**, the other words coming from the auxiliary mode *I*ntroverted *T*hinking *T*i.

The Myers–Briggs Type Table has the same type arrangement as Table 2.3, with type descriptions instead of cognitive modes. With its elegant simplicity, the easy-to-understand Type Table accounts for much of the runaway popularity of the *MBTI*. At the time it was not realized that there were 16 more types unaccounted for and that many people have preferences and skills related to more than two modes. In a sense, the *MBTI* was running at half speed, as will be demonstrated especially in Chap. 4.

Unintentionally perhaps, the Type Table encouraged the notion that there is little variation in personality among people in the same type category. Counterexamples to this plausible but oversimplified idea abound in the rest of the book.

2.4.5 Other Questionnaires

Several other questionnaires covering the same ground have since appeared. Keirsey and Bates (1978) brought out their "Temperament Sorter" with its ten *E–I* and twenty *S–N*, *T–F* and *P–J* items—70 in all. Their response results can be put on a quantitative basis comparable with the 30-item *MBTI* by using the factor 100%/10 = 10% for *E–I* and 100%/20 = 5% for the other three.

With their Steps II and III factor analysis statistical deconstruction of the original *MBTI* instrument, Quenk, Hammer and Majors (2001, 2008) have added items intended to equalize coverage of the four variables. Their quantification factor would be $100\%/5 \times 5 = 4\%$ per item. The Teamology questionnaire used in this section was originally developed for constructing and organizing student design teams, using the twenty Step II "facets".

2.5 Concluding Summary

This chapter has laid out an informal and incomplete history of what might be called the "qualitative era" of Jungian typology. Its intention was to seek out weak spots capable of being strengthened by the quantitative analysis presented in the rest of the book.

Starting with Jung's publication of his *Personality Theory*, contemporary with Briggs' early work with the *P–J* attitude, the chapter sought first to construct an axiomatic basis for the theory. Two postulates seemed complete enough, but the third exposed oversights in the definition and naming of essential attitude variables. But even with one variable unclear, Jung's original theory was shown sufficient for casual estimation of the dominant mode, and, surprisingly, Briggs' supplying of the missing *P–J* attitude did not by itself identify the auxiliary attitude definitely. To do the job, an arbitrary "attitude balance" rule had to be employed. The rule conveniently reduced the number of possible two-mode personality types from 32 down to the manageable sixteen of the famous Type Table.

Even more than the introduction of the four-variable *MBTI*, Myers' simple and well-written Type Table drew international attention to the new qualitative personology. The next chapter sets the stage for its quantitative extension doubling the descriptive power, not only of Jung's underlying personality theory, but also of the *MBTI* instrument itself.

2.6 Chapter 2 Exercises

2-1 This chapter's exercises concentrate on your results on the teamology questionnaire. First, express them as percentages in the *MBTI* order *E–I*, *S–N*, *T–F*, *P–J*.
2-2 What would the *GW* instrument predict as your dominant cognitive mode?
2-3 How would Jung write your type in terms of dominant and auxiliary?
2-4 What would the *GW* instrument predict as your auxiliary function?
2-5 What would Grey and the Wheelwrights predict as your auxiliary attitude?
2-6 What would Myers and Briggs predict as your auxiliary attitude?
2-7 Is your pair of cognitive modes in the Type Table 2.3?
2-8 What, if any, *MBTI* 4-letter type would your cognitive modes generate?

Chapter 3
Quantification

Data, Watson; I must have data!
I can't make bricks without straw!
–Sherlock Holmes in A. Conan Doyle's
"The Sign of Four"

3.1 Introduction

This chapter begins the conversion of Jung's qualitative concepts into a solid and rigorous quantitative theory useful for applications by teamologists and other 8-function(-attitude) personologists. Although finding the right mode categories is all that the applications need, identifying them even qualitatively will involve computing numerical mode scores from MBTI data. The QUANTitative mode scores obtained from the MBTI data will then QUALitatively pinpoint the modes, otherwise known as the Jungian types.

Such use of the numerical data needs to be justified to certified MBTI users, who for good reasons have been schooled to distrust any interpretation of a high MBTI score as suggesting competence or aptitude. It turns out, though, that to identify the Jungian modes associated with any particular set of MBTI type pairs, one must first determine mode scores from their underlying MBTI scores. The mode scores will subsequently indicate the associated Jungian mode types. Thus the numerical MBTI data, which psychologists often describe as "dimensional", is a necessary bridge from the MBTI instrument to Jung's qualitative theory. Section 3.2 therefore seeks to justify the quantification process to come, emphasizing the importance of the MBTI data for ultimately identifying the cognitive modes, the "Jungian types".

The rest of the chapter sets up definitions and conventions for the quantification to follow in Chaps. 4 and 5. The various Jungian instruments differing from the MBTI require some sort of standardization and normalization so that a team, say, can if necessary compare information from different questionnaires when considering team role assignments (Sect. 3.3). Questionnaire uncertainty and personality variability are then considered in Sect. 3.4, where existing MBTI clarity standards are exploited to distinguish significant effects from ambiguously slight ones. The chapter concludes by establishing semi-quantitative symbolic alternatives to the existing sixteen four-letter type tags such as ENTP.

D. J. Wilde, *Jung's Personality Theory Quantified*,
DOI: 10.1007/978-0-85729-100-4_3, © Springer-Verlag London Limited 2011

3.2 Numerical Data

Myers and McCaulley (M&M) have written, "the Indicator is not trying to *measure* people, but to *sort* them into groups to which, in theory, they already belong" (1985, pp 140–141, emphasis in original, abbreviated "Manual2"). In this way M&M distinguish a *trait*, or dimensional degree of preference, from its associated *type*, or preference category. Arnau et al. (2003, p. 234) note that "M&M do not necessarily dispute that individuals possess varying degrees of skill in using each of the attitudes and functions, ...". They (M&M) merely contend that such traits are less important for personality theory than categorical types, a position at least supporting the *existence* of personality traits, if not their relevance.

According to one of their directors, MBTI certification programs seem to have elevated M&M's denial of trait *relevance* into a negation of trait *existence*, indeed teaching prospective practitioners that MBTI "numbers don't measure anything". This unintentional taboo on using numerical data may unfortunately inhibit some readers from even learning about the quantification to follow, which of course must rely on MBTI score data. Jung himself would not hesitate, however, for in *PT* he writes, regarding introverts and extraverts, "There is finally a third group, and here it is hard to say whether the motivation comes chiefly from within or without. This group is the most numerous and includes the less differentiated normal man." (pp 515–516), Thus Jung would think that having a low *E-I* score such as the 20%*E* in Example 1 is *normal*! Why not?

After an extensive empirical study of the MBTI and two related Jungian instruments involving 887 Texas A&M undergraduate and graduate students, Arnau et al. concluded, "Jungian attitude and function preferences appear to be continuous constructs rather than distinct categories". Even the 2005 investigation of MBTI type versus trait by Reynierse and Harker (abbreviated RH), whose "statistical comparisons (of 770 individuals) favored a type interpretation for each of the MBTI preferences", found also that "... the 302 significant within effects indicate the presence of trait effects in these data." Hence both these careful empirical studies support the existence of information in the data after all. Some of this information will turn out to be quite valuable.

Mapping MBTI scores to Jungian types will in fact *require* numerical data, whose numbers are as likely to be small as they are to be large. As pointed out in Sect. 2.4.2, addition of the *P-J* attitude to the *GW* instrument does not by itself determine the auxiliary attitude; a *TD* attitude balance rule is needed. In Chap. 4 it will be proven that this approximate *TD* rule can be replaced by a simple and precise comparison of the *P-J* and *E-I* scores. This new application will demonstrate how TD's qualitative categorical mapping can lose vital typological information.

Rest assured that the quantification to come will not force anyone to interpret MBTI scores as anything other than a means to an end—identifying the preferred cognitive modes. In fact, anyone afraid that Jungian scores might be misused to indicate skill is free to discard them just as soon as the Jungian types have been determined from them.

3.3 Normalization

As noted in Sect. 2.4.5, the MBTI is not the only instrument measuring variables of Jung's personality theory. In order to compare information from different instruments, it is advisable to standardize and normalize their scores. With this done, several questionnaires can be used in any 8 function (-attitude) application, as when team role assignments are being made (Wilde 2009, Chap. 4).

Since normalization was already illustrated in Sect. 2.2.1, its recapitulation here will be brief. Convert any point score to a relative basis by dividing it by the score maximum and multiplying the resulting decimal fraction by 100% to generate a "percentage score", expressed as a numerical percentage followed by the symbol for the variable. In Example 1, a score of 3 for P, whose maximum value is 5, would produce a percentage difference score of (3/5)100% = 60%P. The factor 100% divided by the maximum, here (100%/5) = 20%, can be considered the number of percentage points per questionnaire item. For reference, this multiplier will be known as the *normalization factor*. Thus one could generate the final percentage score by allotting 20% to each item so that the single J response (20%J) subtracted from the sum of the four P responses (80%P) would yield the total difference desired: 60%P.

A more substantial situation, involving the complete MBTI for the same person and designated "Example 1b", has absolute scores of 7E, 28N, 4T, 16P with 30 items per variable. The reader can verify that the corresponding percentage scores are 23%E, 93%N, 13%T, 53%P, where these results have been rounded to avoid fractions and decimals. In qualitative MBTI terminology this is known as an *ENTP* "whole type".

Opposite variables are given opposite signs. Thus the example results can also be expressed as −23%I, −93%S, −13%F, −53%J. This is needed whenever the formulas to be developed in what follows involve scores for variables opposite to the positive ones measured.

Normalization may not be strictly needed by anyone working with a single questionnaire to which no comparisons are to be made. Even then, however, normalization will turn out to simplify greatly the operations required for decoupling the attitudes and forming the cognitive mode scores.

3.4 Clarity and Significance

Page 121 of Manual3 (the abbreviation for Myers et al. 1998) asserts, "The number associated with an MBTI preference … is most appropriately interpreted as providing information about the likelihood that the preference has been correctly reported". This speaks to uncertainty, both in the instrument itself and in the person filling it out. For this reason what had previously been called "scores" were renamed "preference clarity indices (pci's)" or "clarities" in Manual3. This

Table 3.1 MBTI score and percentage descriptors

pci	% Range	Descriptor
0–5	0–19	Slightly clear
6–11	20–39	Moderately clear
12–23	40–79	Clear
24–30	80–100	Very clear

distinction will be followed only casually here because the book usually focusses on mode identification rather than clarity.

For people uncomfortable with numbers or percentages, Manual3 further advises. "An alternative to reporting the pci itself is to present the preference clarity index category ("slight", "moderate", "clear", "very clear") to the respondent instead." Table 3.1 gives the pci and percentage ranges for these descriptors (ibid, pp 122–123).

To simplify things further, all scores 20% or higher will henceforth be considered "definite" or "significant", whereas smaller ones will be regarded as "slight", e.g., 13%*T* in example 1b. In practice, single-digit % scores will sometimes be considered "ambiguous" or neglected entirely.

3.5 Non-Numerical Semi-Quantification

Manual3's approach in the preceding section to data clarity could be characterized as a "non-numerical semi-quantification", in that it replaces the pci quantitative data with verbal descriptors associated with number ranges rather than individual scores. This compromise is worth extending to the four letter pairs of the MBTI type tag itself, although this concept should certainly be regarded as optional in practice.

The idea is simply to rearrange the letters in descending score order. Thus in ENTP example 1b the new "ranked" type tag would read "NPET". It does not take much imagination or type experience to realize that an NPET brain stormer would differ noticeably from someone having the reverse order and more directive TEPN on his lapel. Such interesting differences within the same whole type, here ENTP, will be brought out more precisely and dramatically once the full quantitative theory has been developed in future chapters. For the moment, consider it an interesting temporary notion, entirely optional.

This ranking idea was inspired by the 2005 research of RH referenced in Sect. 3.2. Ranking the variables in this way for each sample item, they were able to observe the highly significant type and trait effects they reported.

At the risk of trifling over-complexity, consider also combining Manual3's semi-quantitative clarity approach with significance simplification. That is, express the **clear** and **very clear** variables in **boldface** and display the slight variables in plain lower case. Example 1b would then give ***NPEt*** for the MBTI results. It may be interesting that Example 1a involving the coarser teamology questionnaire of

Chap. 2 would yield the slightly different result ***NP****(et)*, the parentheses indicating the tie between 20%*E* and 20%*T* for the reduced number of questions.

Although this semi-quantitative approach certainly has immediate application, it should be considered merely a temporary suggestion, for clearer representations will appear after the attitudes have been decoupled in the next chapter. The mode representations of Chap. 5 and beyond will be even more suggestive and powerful.

3.6 Concluding Summary

This short chapter sets the stage for the quantification to come of Jung's personality theory. It expends considerable effort to convince wary graduates of MBTI certification programs that the numerical data, although perhaps not measuring competence in a preference, does indeed contain quantitative information essential for indicating the correct cognitive modes of Jungian theory. It lays out a straightforward normalization procedure allowing objective comparison of the various Jungian instruments and preparing the ground for amazingly simple formulas to be derived in Chaps. 4 and 5 and beyond. It then normalizes the Manual3 approach to instrument variability, which interprets the data in terms of preference clarity. This informal verbalization of the quantitative data is then extended to the letter indicators themselves, following a ranking strategy of Reynierse and Harker. Informative though this new representation may be, it will soon be superseded by more precise and comprehensive quantifications in the chapters to follow.

3.7 Exercises

3-1 Express your results on the teamology questionnaire semi-quantitatively as a ranked type tag in the manner of Sect. 3.5.

3-2 If you have MBTI scores, express them semi-quantitatively as a ranked type tag in the manner of Sect. 3.5.

3-3 Do the same for Example 2 to come in Ch. 4. The attitude scores are reversed from Example 1 so that Example 2 is 60%*E*, 20%*P*, 100%*N*, 20%*T*.

3-4 Do the same for Example 3 to come in Sect. 6.3: 100%*E*, 100%*P*, 40%*N*, 20%*T*.

3-5 Do the same for Example 5 to come in Sect. 6.5: 100%*I*, 100%*S*, 100%*T*, 100%*J*.

3-6 Do the same for Example 6 to come in Sect. 6.6: 100%*E*, 60%*J*, 60%*S*, 20%*F*.

Chapter 4
Decoupling the Attitudes

Nothing short of independence can possibly do.
–George Washington, April 21, 1778

4.1 Introduction

This chapter shows how to decouple the attitudes, an activity whose procedure or even purpose may not be at all clear to most readers. It is therefore essential to explain here not only what is meant by "coupling" but also why undoing it is well worth the effort.

Consider the two sets of cognitive modes conceived by Jung, one for perception (information gathering) and the other for judgment (decision-making). As stated in Postulate 3 of Sect. 2.2.3 in Chap. 2, the perception modes involve only the perception functions; the judgment modes, only the judgment functions. Thus Jung conceived the two mode sets, henceforth called "domains", as completely *independent* of each other as far as the psychological functions are concerned. That is, he saw the *p*-modes as unaffected by the *j*-functions *T* and *F*, just as the *S*–*N* *p*-function pair has no influence whatever on the *j*-domain.

Unfortunately, this independence doesn't hold for the traditional psychological *attitude* pairs *E*–*I* and *P*–*J* because they always affect both domains, as portrayed in Fig. 4.1a. Hence the two domains are unfortunately "coupled" by these attitudes, a situation in conflict with Jung's conception, expressed in Postulate 3, of attitudes that are mutually independent, one pair affecting only the *p*-domain and the other pair only the *j*-domain. That's why Sect. 2.2.3 in Chap. 2 formulated a different couple of attitude pairs: extraverted and introverted perception *E*p–*I*p for the *p*-domain, and similarly, extraverted and introverted judgment *E*j–*I*j for the *j*-domain. These new attitudes are said to be "decoupled" because each only affects its own domain. This makes the two domains truly independent with respect to the new *decoupled* attitudes, as pictured in Fig. 4.1b. Since the domains are also independent with respect to the psychological functions, they can then be regarded as mutually independent in all respects after attitude decoupling.

This importantly advances Jung's personality theory because it also *simplifies* it. Without this decoupling, all four variable pairs must always be considered simultaneously, making for a four-dimensional problem difficult if not impossible

D. J. Wilde, *Jung's Personality Theory Quantified*,
DOI: 10.1007/978-0-85729-100-4_4,

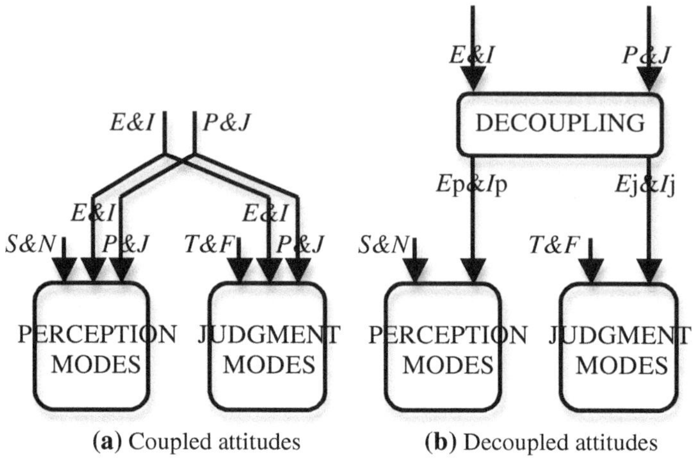

Fig. 4.1 Attitude decoupling

even to visualize. But after decoupling, instead of a single problem of four dimensions there remain two entirely separate sub-problems, each having only two dimensions. Even though this leaves two sub-problems instead of one, each can be easily visualized on an ordinary Cartesian graph. As will be developed in Chap. 5, each of these domain graphs will simply involve the domain function score on the horizontal (x-) axis and the decoupled domain attitude score on the vertical (y-) axis.

The remaining question then is how to convert the traditional *E–I* and *P–J* data into scores for the new decoupled attitude pairs *E*p–*I*p and *E*j–*I*j. Section 4.3 will accomplish this by employing a style of reasoning dating back to the classic Greek philosopher Plato. In the present application, Plato's concept of "ideal" rigorously suggests matching any maximum possible 100% decoupled attitude score with equivalent 100% scores of the associated original *MBTI* attitudes. Geometrically speaking, this generates a new coordinate system with axes rotated 45° from the original ones. In this new system it is easy to score the more usual non-ideal data automatically by interpolation. The additive formulas for this are as simple as one could hope—mere averages of sums or differences of the original scores. The "decoupling" box in Fig. 4.1b symbolically depicts this conversion of coupled attitudes into decoupled ones.

Readers intimidated by mathematical reasoning would be justified in skipping directly to the very simple results of Sects. 4.3.2 and 4.3.3. Anyone is free to accept the plausible results on faith or rely on their more skeptical colleagues to verify correctness. This is in fact a reading strategy recommended to eight-function (-attitude) consultants wishing to get on with solving their clients' problems immediately rather than to stop and check the underlying reasoning.

To obtain the score of the principal (larger) *de*coupled attitude, merely average the positive *E–I* and *P–J* scores. That is, divide their sum by two to allow for there originally being two variables. The capital letter for this principal *de*coupled attitude will be the same as the *E–I* letter—*E*s have either *E*p or *E*j; *I*s, either *I*p or *I*j. The principal decoupled attitude domain is the *same* as the *P–J* letter for *E*-people: *E*p for *P*s and *E*j for *J*s. However, this situation is *reversed* for *I*s; *I*j for *P*s and *I*p for *J*s. This confusion, which results from the misleading way *P* and *J* were originally named (Sect. 2.4.3 in Chap. 2), will disappear once the new decoupled attitudes are defined.

After the principal decoupled attitude has been identified and scored, it is easy to find the "subsidiary" decoupled attitude in the other domain, whose subscript is of course already known. Its score is simply the difference between the scores of the larger coupled attitude and the principal attitude just calculated. When the larger score is for *E–I*, the subsidiary capital letter will be the *same* as for the principal attitude. Otherwise, as for most people, the subsidiary capital will be *opposite* to that for the principal attitude.

These decoupled attitudes deserve as much attention as the traditional coupled ones have received, and indeed, they might bring a fresh look at personality theory. To this end the new attitudes are named in Sect. 4.8, which then furnishes short descriptions of them useful for informal analysis.

Replacing the old attitudes with the new ones forces re-examination of the *MBTI* letter code in Sect. 4.9. It is natural to group each decoupled attitude with its function in the same domain, a procedure to be simplified even further when Chap. 5 shows how to compute the corresponding modes. Thus the results of Chap. 4 become the inputs to Chap. 5 where the main results of the book are presented.

4.2 Attitude Transformation

Both traditional attitude pairs *E–I* and *P–J* are inputs to both the perception domain and the judgment domain *simultaneously*. In other words, changing any traditional attitude affects EVERY mode score. For this reason, these attitudes and both domains are said to be "coupled".

This conflicts with Jung's conception of two domains completely independent of each other as formulated in Postulate 3. Consequently these coupled attitudes must be replaced by the two new "decoupled" attitude pairs *E*p–*I*p and *E*j–*I*j introduced in Sect. 2.2.3 in Chap. 2 that dealt with Postulate 3.

The task of the present section is to develop the decoupling operation, symbolized in Fig. 4.1b by the box labeled "DECOUPLING". Since there are only two variables to be transformed, the problem has but two dimensions and so can be depicted graphically as in Fig. 4.2 in which *E–I* is plotted against *P–J*. In mathematical language, the decoupling problem has "rank 2", in contrast with the original rank 4 problem.

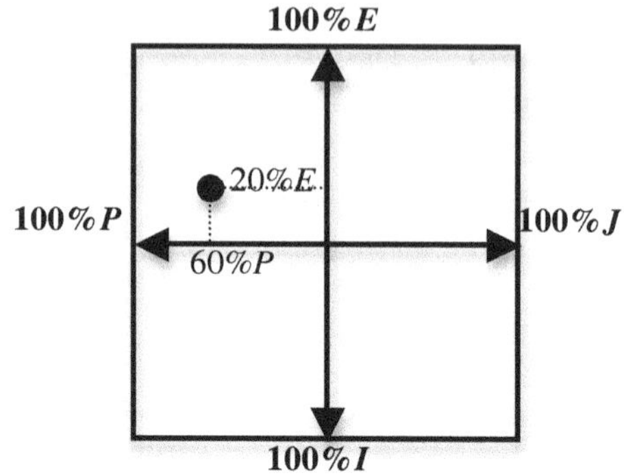

Fig. 4.2 Coupled attitude space

4.2.1 Bidirectional Axes

Notice that the coordinate axes do not follow the usual Cartesian convention in which each unidirectional axis has a negative as well as a positive section. The present application instead has bidirectional axes radiating outward from the origin where the axes intersect. With this arrangement all variables are positive *always.*

Thus the *E*-axis points upward from the origin; the *I*-axis, downward. Similarly the *P*-axis points to the left; the *J*-axis, to the right. This deviation from mathematical convention is a concession to the preference by many psychologists for positive numbers over negative ones. The prefix "100%" on each variable is a reminder that the entire range is being used for every variable. The origin is of course the location of 0% for both variables. To illustrate the new sign conventions, Example 1a attitude scores 20%*E* and 60%*P* are plotted as a point in Fig. 4.2.

In similar fashion the two output variables, also a rank 2 system, can be represented graphically in Fig. 4.3 as a plot of the decoupled variables *E*p–*I*p vs. *E*j–*I*j, even though it is not yet clear how to find them from the coupled ones. Anticipating the soon to be derived solution, the coordinate axes are drawn between opposite corners of the square instead of in the customary horizontal and

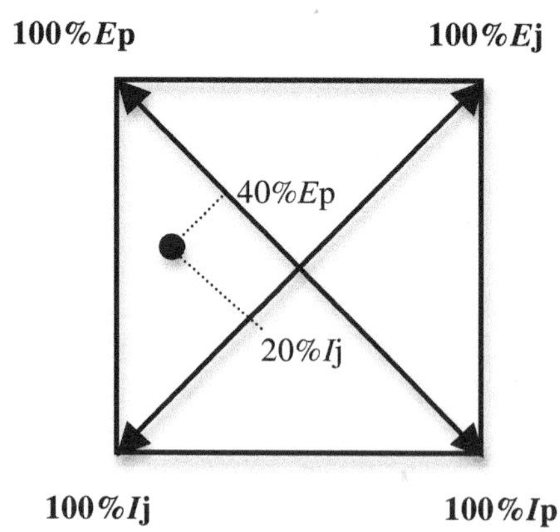

Fig. 4.3 Decoupled attitude space

vertical directions. To show how to read the new "diagonal" coordinates, the coordinates 40%*E*p and 20%*I*j, whose significance will be revealed later, are plotted as a point. The corners of the two graphs have been matched deliberately so that both cover the same set of points.

4.2.2 Opposite Attitudes, Opposite Signs

Eventually the attitude transformations will be expressed as simple formulas such as $Ep = (E + P)/2$. What is to be done if the original data are not in terms of the formula variables? Suppose for example that the *MBTI* attitudes are 30%*I* and 50%*J*, both variables opposite to those in the formula. Deal with this by making the signs of the opposite variables *negative* while reversing the letter symbol to match the formula. Thus 30%*I* becomes −30%*E*, and 50%*J* is written −50%*P* in the formula, which becomes $Ep = (E + P)/2 = (-30\%E - 50\%P)/2 = -40\%Ep = 40\%Ip$. Here the negative value −40% for the decoupled attitude *E*p is taken care of by substituting the opposite decoupled attitude *I*p. In this manner one always can begin and end with positive numbers, using negative signs to indicate opposite variables, and vice versa.

All that is needed now is to establish a correspondence, or "mapping", between three points in Fig. 4.2 and three points in the same locations in Fig. 4.3. The only geometric restriction is that the points in each trio not be in a straight line. As a preview, the points matched will be the origins and the two upper corners in each graph.

4.3 Platonic Ideal Attitudes

This section justifies the point matching just mentioned in the section preceding. The line of reasoning goes back 2,500 years to Plato's famous work *The Republic*, in which he used the concept of the "ειδοσ", the "ideal", to examine the order and character of the City–State. In the context of the present book, "ideal" *de*coupled attitudes will be expressed as combinations of "ideal" coupled attitudes in order to establish numerical correspondences between them.

Here the numerical concept of "maximum" will take the place of Plato's philosophical concept of "ideal". For instance, 100%*E* will be considered the score for an ideal extravert, and similarly for the other three coupled attitudes and four decoupled attitudes as detailed in the subsections following.

4.3.1 Platonic (Ideal) Reasoning

To construct the desired Platonic correspondences between coupled and decoupled attitudes, it helps to review the verbal interpretations of the Myers–Briggs

P–J attitudes, which until now have been dealt with entirely as symbols. A widely accepted interpretation of "pea", readily inferred from the five questions of Sect. 2.4.3 in Chap. 2, is that it describes a free-and-easy flexible attitude contrasting with the structured "jay" approach. Thus combining *E*xtraversion with *P*'s flexibility produces the Extraverted Perception attitude *E*p, here labeled "Exploration". Numerically and Platonically then, 100%*E*xtraversion combined with 100%*P* flexibility is considered to generate a 100%*E*p exploration attitude.

Opposite to *E*p is the introverted perception attitude *I*p, for which the "Focus" Platonic 100%*I*p (= −100%*E*p) attitude corresponds to the Platonic coupled attitudes 100%*I*ntroversion and 100%*J* (structure). But this is exactly the relation obtained by reversing all signs in the extraverted perception expression: 100%*I*p = −100%*E*p = (−100%*E* − 100%*P*)/2 = (100%*I* + *100%J*)/2.

There is danger here of unintended confusion because this introverted *perception* formula depends on the coupled attitude "jay" rather than on "pea". Indeed, if Myers and Briggs' original terminology "judgment" and "perception" had been employed for "jay" and "pea", the formula would give the contradictory impression that somehow a decoupled perception attitude depends on a coupled "judgment" attitude, which of course would be semantically absurd. As mentioned in Sect. 2.4.3 in Chap. 2, it was to avoid such a categorical error that these misleading names were abandoned in favor of the abstract letter names "pea" and "jay". Beware of such categorical confusion whenever an attitude is introverted for, as will be discussed in Sect. 7.3 in Chap. 7, *M* and *B*'s verbal terminology, good enough for extraverted attitudes and modes, is completely backwards when they are introverted.

To continue the Platonic correspondences, maximum *E*xtraversion combines with maximum structure *J* to yield maximum extraverted judgment *E*j, here dubbed "Control". Just change signs to reveal the maximum introverted judgment attitude *I*j, named "Appraisal" here, as corresponding to maximum *I*ntroversion and flexibility *P*. But stay alert to avoid categorical confusion, which fortunately will leave the scene once the *P–J* pair has been replaced by the new decoupled attitudes.

4.3.2 Quantification in the Perception Domain

The time has finally come to quantify all this qualitative reasoning and quantitative preliminaries. Starting in the perception domain, there is only one way to combine the two Platonic maximum coupled attitude scores 100%*E* with 100%*P* to obtain the maximum Platonic decoupled attitude score of 100%*E*p. That is to normalize by dividing the sum (200%) of the coupled scores by 2 to obtain 100%*E*p = (100%*E* + 100%*P*)/2. This mean or average sum exactly maps the upper left corner (100%*P*, 100%*E*) of coupled attitude graph (Fig. 4.2) to the upper left corner (0%*E*j, 100%*E*p) of decoupled attitude graph (Fig. 4.3).

What is needed however is an equation usable throughout the coupled and decoupled spaces in the more usual non-ideal situations. This is easily obtained by removing the 100% coefficients to obtain

$$Ep = (E + P)/2 \tag{4.1e}$$

Equation 4.1 is valid for both entire graphs (Figs. 4.2, 4.3). The graphs will then match exactly not only at their upper left corners (100%*P*, 100%*E*) and (100%*E*p, 0%*E*j), but also at their origins (0, 0) and lower right corners (100%*J*, 100%*I*) and (100%*I*p, 0%*E*j). Equation 4.1e amounts to an interpolation formula for general non-ideal percentages. For the values 20%*E* and 60%*P* of Example 1, Eq. 4.1 gives (20 + 60)%/2 = 40%*E*p, the decoupled *p*-domain score of Sect. 4.2.1 and Fig. 4.3.

Be careful when working with introverted attitudes. To see what can happen, consider the following equation obtained by multiplying Eq. 4.1e throughout by −1 and introducing the needed opposite variables to preserve positivity:

$$Ip = (I + J)/2 \tag{4.1i}$$

The *I* is certainly reasonable in an introverted situation, but having *J* in a perception attitude formula may at first seem counter-intuitive. This is an unfortunate artifact of the misleading way the *M–B* attitudes were labeled. As noted in 2009 by Reynierse, this comes from the categorical error in logic discussed in Sect. 7.3 in Chap. 7. It is worth observing that in the perception domain, coupled attitudes will always combine either as $E + P$ or as $I + J$ (= $-(E + P)$) in the transformation formulas.

4.3.3 *Quantification in the Judgment Domain*

The same reasoning holds for extraverted judgment *E*j on the other decoupled axis, which runs from the upper right corner (100%*E*j, 0%*E*p) of Fig. 4.3, through the origin (0, 0), to the lower left corner (100%*I*j, 0%*E*p). The resulting formula for *E*j involves *E* again, this time averaged with the other Briggs attitude *J* instead of with *P* as before.

$$Ej = (E + J)/2 \tag{4.2e}$$

It is easy to verify that this *mean sum* relation matches the Platonic ideal coupled attitudes 100%*E* and 100%*J* with the ideal decoupled attitude 100%*E*j. Equation 4.1 also shows that the perception decoupled attitude vanishes as required, for (100%*E* + 100%*P*)/2 = (100%*E* − 100%*J*)/2 = 0%*E*p.

As in the *p*-domain, the introverted form

$$Ij = (I + P)/2 \tag{4.2i}$$

may seem counter-intuititive because of the appearance of *P* in a judgment domain formula. But this is just another artifact of the misnaming of the *M–B* attitudes

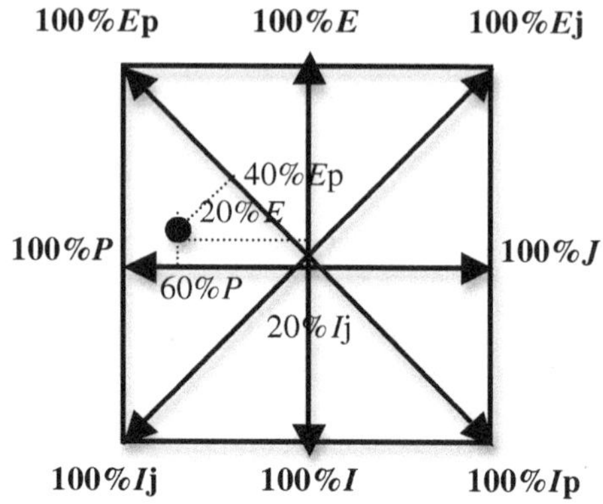

Fig. 4.4 Superimposed attitude spaces

discussed earlier. Notice that in the *j*-domain, attitudes always combine either as $E + J$ or as $I + P$ $(= -(E + J))$.

The new attitude pairs *E*p–*I*p and *E*j–*I*j are said to be "decoupled" because they operate independently only in their own domains, perception and judgment respectively. That is, changing a decoupled attitude affects only its particular domain, as Jung intended.

Equations 4.1 and 4.2 imply that Figs. 4.2 and 4.3 can be superimposed as in Fig. 4.4 following. It should come as no surprise that the points of Example 1 occupy the same locations in Figs. 4.2 and 4.3. This graphical interpretation is developed further in Sect. 4.6.

4.3.4 Platonic Idealization of the Coupled Attitudes

Platonic idealization of the original coupled attitudes can unlock some of the mystery long associated with them. First imagine a Platonic ideal extrovert with coupled attitude scores 100%*E* and 0%*P* = 0%*J*, the latter to isolate the *E* effect. Equations 4.1e and 4.2e then give decoupled attitude scores of 50%*E*p and 50%*E*j respectively, which seems fair enough once one accepts the existence of double extraversion. Similarly the ideal Introvert 100%*I*, 0%*P*(*J*) would by Eqs. 4.1i and 4.2i have its total introversion partitioned equally between the *p*- and *j*-domains: 50%*I*p and 50%*I*j.

Even more interesting perhaps are the Platonic ideal flexible *P* and structured *J* personalities, which would have zero total extra (intro) version 0%*E*(*I*) to isolate the *P*–*J* effects. The ideal flexible personality 0%*E*, 100%*P* would by Eqs. 4.1e and 4.2i have positive decoupled attitude scores of 50%*E*p and 50%*I*j, allocating extraversion to the *p*-modes and introversion to the *j*-modes. Similarly the ideal structured personality 0%*E*, 100%*J* would by Eqs. 4.1i and 4.2e have positive decoupled attitude scores of 50%*I*p and 50%*E*j.

4.4 Mean Sum and Mean Difference

Example 1 does not immediately appear to apply in the judgment domain because Eq. 4.2 involves only *E* and *P* whereas Eq. 4.2i involves only *I* and *P*. To see that it really does apply, express the positive attitude with the smaller score, 20%*E* (<60%*P* in the example), in terms of its opposite score −20%*I*. In this way using the introverted expression 4.2i in Example 1a yields *I*j = (−20%*E* + 60%*P*)/2 = 20%*I*j, which is of course also positive. The decoupled solution for the example is therefore confirmed to be 40%*E*p and 20%*I*j.

The quantity calculated is known as the *mean difference* of the original (non-negative) coupled attitudes. This procedure will always produce a non-negative decoupled attitude (sometimes zero) in both domains.

To emphasize the effect of coupled attitude score size, consider the following Example 2 in which the attitude scores are reversed, i. e., 60%*E* and 20%*P*. This time the *E–I* score is larger, whereas in Example 1 it was the *P–J* score. The mean sum and mean difference are 40 and 20% as in Example 1, but although the mean sum is associated with the same decoupled attitude as in Example 1, (60%*E* + 20%*P*)/2 = 40%*E*p), now the mean difference is associated with a different decoupled attitude (60%*E* + 20%(−*P*))/2 = (60%*E* − 20%*J*)/2 = 20%*E*j, extraverted this time instead of introverted as in Example 1. Some readers may object that having two extraverted attitudes violates the attitude balance principle of type dynamics mentioned in Sect. 2.4.2 in Chap. 2. Realize however that this "principle" is merely an approximion that is not always valid, a topic to be discussed thoroughly in Sect. 7.3.3 in Chap. 7.

4.5 Unpacking

In any particular case, one of the four Eqs. 4.1 or 4.2 will give the largest decoupled attitude score. Let this largest score be symbolized by *D* and called "the score of the *principal* decoupled attitude". The adjective "principal" reflects that *D* will always be at least as large as, and usually larger than, the score of the other "*subsidiary*" decoupled attitude, symbolized by *d*, in the opposite domain.

To avoid treating the four cases of Eqs. 4.1 and 4.2 separately, let *L* be the larger and Σ the smaller coupled attitude scores, with equality permitted ($L \geq \Sigma$). To avoid abbreviating "smaller" by "*S*", a letter reserved for "sensing", the Greek letter "Σ" (capital sigma) is used. Then the principal attitude score is given by

$$D = (L + \Sigma)/2 \tag{4.3}$$

Of course, *D* can be associated with either domain. For Example 1's *L* = 60%*P* and Σ = 20%*E* this would produce *D* = 40%*E*p.

In the other domain, the preceding section showed that the score *d* of the subsidiary decoupled attitude is determined by changing the sign of the smaller decoupled attitude. Thus

$$d = (L - \Sigma)/2 \tag{4.4}$$

Another expression for d is obtained by adding and subtracting $L/2$:

$$d = (L - \Sigma)/2 = L/2 - \Sigma/2 + (L/2 - L/2) = L - (L + \Sigma)/2 = L - D$$

or

$$d = L - D \tag{4.5}$$

This will be known as the "attitude unpacking relation" because it treats the larger coupled variable L as a sum, or "packing", of the principal and subsidiary decoupled attitudes. Thus one can say that the subsidiary attitude is obtained by "unpacking" the larger coupled attitude L, that is, removing (subtracting) the principal decoupled mode D from it. This approach is perhaps easier than fiddling with the signs and dividing by two again. In Example 1, where $L = 60\%P$ and $D = 40\%E$p, this would give the difference 20% as the subsidiary score in the j-domain, either Ej or Ij as determined in the next paragraph.

In general it remains to determine whether the subsidiary decoupled attitude is introverted or extraverted. If L is either P or J as in Example 1, changing the smaller coupled attitude from E to I or from I to E as required by Eqs. 4.1 and 4.2 would generate a similar change in the coupled attitude capital letter in the subsidiary domain. In Example 1 this change produces 20%Ij for the subsidiary.

If on the other hand L is either E or I, changing the smaller coupled attitude from P to J or from J to P as required by Eqs. 4.1 and 4.2 would not alter the coupled attitude capital letter in the subsidiary domain. Thus in Example 2 of the preceding section, in which $L = 60\%E$ and $S = 20\%P$, the unpacked score, still 20%, is associated with 20%Ej in contrast with the 20%Ij of Example 1.

Table 4.1 summarizes the developments of the two sections preceding. The two rows represent total extraversion E and total introversion I; the two columns, the M–B attitude pair P and J. The four cells contain the decoupled attitude formulas as functions of the coupled attitudes. Keywords describing each decoupled attitude and discussed in Sect. 4.8 have been added.

Table 4.1 Coupled and decoupled attitudes

	P	J
E	Extraverted Perception Ep = $(E + P)/2$ EXPLORATION	Extraverted Judgment Ej = $(E + J)/2$ CONTROL
I	Introverted Judgment Ij = $(I + P)/2$ APPRAISAL	Introverted Perception Ip = $(I + J)/2$ FOCUS

4.6 Graphical Interpretation

The two-dimensional attitude space portrayed in Fig. 4.5 following has two coordinate systems. Expressed in unidimensional Cartesian terms, they are (1) *%J* vs. *%E,* with orthogonal axes at right angles, and (2) *%Ep* vs. *%Ej,* with axes still at right angles but rotated 45^0. This situation lends itself naturally to graphical analysis, with accompanying visual understanding and computational insight.

The rectangular coordinates run horizontally from *P* on the left to *J* on the right, and upward from *I* to *E*. The diagonal coordinates *E*p–*I*p run from lower right to upper left; *E*j–*I*j, from lower left to upper right. The region of slight scores (<20%) is shown as a square with a dashed boundary surrounding the origin. Coupled attitudes are plotted in the rectangular coordinates, whereas decoupled attitudes are plotted diagonally. Rectangular coupled points such as (60%*P*, 20%*E*) can be projected 45° on to the diagonal coordinates to determine the corresponding decoupled attitude values such as the (20%*I*j, 40%*E*p) in Fig. 4.5.

In both examples, the mean sum is 40%*E*p associated with *E* and *P*. The mean differences have the same value (20%) but point in opposite directions. Thus in Fig. 4.5, *d* is 20%*I*j, which is shown as a dark point with associated lines. On the other hand, Example 2 has 60%*E* > 20%*P*, implying that this time the diagonal judgment coordinate *d* is 20%*E*j. It is shown as a light point with the associated lines shown.

4.7 Doubleversion

This section is merely a note on terminology. As in Example 1, most people have one decoupled attitude extraverted and the other introverted. But as in Example 2,

Fig. 4.5 Coupled and decoupled attitudes

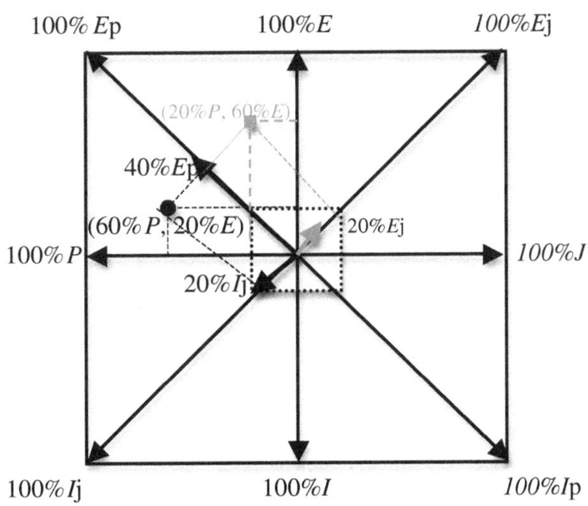

Example 1 shown dark; example 2 shown light.

there are personalities in which both decoupled attitudes are extraverted, who will be referred to in this book as "double extraverts". Similarly, there exist double introverts for whom both decoupled attitudes are introverted. Double extraverts and double introverts will collectively be known as "doubleverts", a group shown in Chap. 7 to be unrecognized by the traditional methods of Type Dynamics.

4.8 Recoupling Inversion

Coupled attitude scores can be recovered from known decoupled scores by adding or subtracting the defining equations. Adding the decoupled *E*p and *E*j equations and canceling out $P + J = 0$ gives the *re*coupling equation for *E*xtraversion

$$E = E\mathrm{p} + E\mathrm{j} \tag{4.6E}$$

Multiplication throughout by -1 yields the recoupling formula for *I*ntroversion:

$$I = I\mathrm{p} + I\mathrm{j} \tag{4.6I}$$

Thus Jung's Postulate 1 extraversion score is the algebraic sum of the extraversion scores from the two domains. Thus *E–I* may be considered *net* extra(intro)version.

Subtraction of the *E*j equation from the *E*p equation gives the interesting result

$$P = E\mathrm{p} - E\mathrm{j} = E\mathrm{p} + I\mathrm{j} \tag{4.7P}$$

The flexibility score *P* is the difference of the extraverted decoupled *p*-attitude and the *j*-attitude. Alternatively, it is the sum of the *p*-domain's *extraverted* score and the *j*-domain's *introverted* score. In terms of the new decoupled attitude keywords, flexibility combines *p*-domain exploration with *j*-domain appraisal.

Multiplication throughout by -1 yields an equally interesting formula for the structured attitude score *J*:

$$J = I\mathrm{p} - I\mathrm{j} = I\mathrm{p} + E\mathrm{j} \tag{4.7J}$$

Thus *p*-domain introversion and *j*-domain extraversion add up to make the structured attitude *J* score. Structure then is made up of both *p*-domain focus and *j*-domain control. Thus *P–J* may be considered *directed* extra(intro)version.

4.9 Naming and Describing the Decoupled Attitudes

The newly decoupled attitudes deserve at least as much attention as the traditional coupled attitudes for which there is already an extensive literature. The keywords proposed in Table 4.1, intended to be provisional names, are a first step in this direction. Following is a brief discussion of each decoupled attitude to lay the groundwork for expanded descriptions by experienced psychologists and typologists as they realize the importance of attitude decoupling. The descriptions here combine those of the various facets of the *E–I* and *P–J* attitudes shown on p. 5 and p. 9 of the *MBTI Step II Manual* (Quenk, Hammer and Majors). They are certainly not presented as the last word on the decoupled attitudes; experienced personologists are welcome to improve both their expression and their content.

4.9.1 Extraverted Perception: Exploration

Extraverted perception *E*p involves externally oriented expressive open-ended improvisation.

Such an exploring person: Connects well with people. Values newness and variety. Makes interests obvious. Plunges ahead without detailed plans. Enjoys meeting new people. Keeps options open. Learns actively by doing, listening and questioning. Works well against a deadline. Is talkative, lively and flexible.

4.9.2 Introverted Perception: Focus

Introverted perception Ip involves interior contained planned structure.

Such a focused person: Does not initiate talk with relative strangers. Is comfortable with routine. Appears difficult to know. Is good at following directions. Prefers talking individually with people known well. Likes order. Learns reflectively. Starts projects early. Is quiet and reserved. Values planning.

4.9.3 Extraverted Judgment: Control

Extraverted judgment *E*j involves externally oriented expressive methodical structure.

Such a controlling person: Connects well with people. Is comfortable with routine. Makes interests obvious. Is good at following directions. Enjoys meeting new people. Likes order. Learns actively by doing, listening and questioning. Starts projects early. Is talkative and lively. Values planning.

4.9.4 Introverted Judgment: Appraisal

Introverted judgment *I*j involves interior reserved improvisational variety.

Such an appraising person: Does not initiate talk with relative strangers. Values newness and variety. Appears difficult to know. Plunges ahead without detailed plans. Prefers talking individually with people known well. Keeps options open. Learns reflectively. Works well against a deadline. Is quiet, reserved and flexible.

4.10 Decoupled Representation

Decoupling of the attitudes suggests a new way of representing a numerical type that replaces the coupled attitude pairs *E–I* and *P–J* with the coupled attitude pairs *E*p–*I*p and *E*j–*I*j. Thus in Example 1, with *MBTI* data 20%*E,* 100%*N*, 20%*T*, 60% whose categorical representation is *ENTP, t*he normalized decoupled representation might be written 40%*E*p, 100%*N*; 20%*I*j, 20%*T.*

Following Manual3's non-numerical semi-quantification idea discussed in Sect. 3.5 in Chap. 3, the letters for the decoupled variables could also be arranged in descending score order without explicitly showing the scores. For Example 1 this would produce the decoupled and ordered letter code *NE*p *I*j*T*, admittedly somewhat complicated and hard to read even with the space introduced to separate the perception from the judgment variables. As suggested in Sect. 3.5 in Chap. 3, boldface and lower case can be introduced to emphasize numerical differences in a manner that is only semi-quantitative: ***NE*p** *ijt*. This grouping, suggested here only tentatively, anticipates representations of Jung's cognitive modes to be discussed in Chap. 5 and 6.

4.11 Concluding Summary

After some words to persuade traditional—even skeptical—personologists to read into the mildly radical innovations of attitude decoupling, the chapter provided informal but rigorous proofs for theorists to verify. Remarkably, the overall problem is reduced from one with an unwieldy four dimensions to two smaller ones, each tractably having only two dimensions. The simple mean sum and difference formulas resulting can be immediately put to work by problem solvers: 8-function (-attitude) practitioners, teamologists and Jungian analysts. A simpler alternative is to "unpack" the larger coupled attitude, obtaining the subsidiary decoupled attitude by subtracting the mean sum principal decoupled attitude from the larger coupled attitude. Accompanying graphics showed that the decoupling transformation is equivalent to a straightforward 45 degree rotation of the coordinate axes. Decoupled attitude descriptions were then provided to encourage extension of the typological approach, already applied extensively to the functions, to the heretofore less studied attitudes.

This sets the stage for the important mode score computations of the next chapter. Fortunately for the reader who has mastered the present chapter, the theoretical developments to follow have exactly the same form as those used to decouple the attitudes. Thus the hard work is already behind!

4.12 Exercises

4-1 Calculate the decoupled attitudes from your results on the teamology questionnaire.
4-2 If you have *MBTI* scores, calculate your decoupled attitudes from them.
4-3 Do the same for Example 2 to verify the results in Sect. 4.4: 60%*E*, 20%*P*, 100%*N*, 20%*T*.
4-4 Calculate the decoupled attitudes for Example 3 to come in Sect. 6.3 in Chap. 6: 100%*E*, 100%*P*, 40%*N*, 20%*T*.
4-5 Calculate the decoupled attitudes for Example 5 to come in Sect. 6.5 in Chap. 6: 100%*I*, 100%*S*, 100%*T*, 100%*J*.
4-6 Calculate the decoupled attitudes for Example 6 to come in Sect. 6.6 in Chap. 6: 100%*E*, 60%*J*, 60%*S*, 20%*F*.
4-7 For any of the preceding exercises, plot coupled and decoupled scores. Also write the decoupled ordered type code in the manner of Sect. 4.9.

Chapter 5
Cognitive Mode Determination

A theory should be simple, but not too simple.
–Albert Einstein

5.1 Introduction

Dedicated readers who have mastered the decoupling formulas in Chap. 4 may still be wondering why it is worth the effort. The present chapter will show that decoupled attitudes make for simpler expressions when they are combined with the functions to form cognitive modes.

The preceding chapter showed how to implement Jung's Postulate 3 concept of two entirely independent cognitive mode domains. The two domains are completely independent of each other because the decoupled attitudes Ep and Ej have been constructed to be independent. This reduces the original 4-dimensional problem down to two 2-dimensional problems, an important and completely rigorous simplification.

In the present chapter these attitude and function pairs are combined into four pairs of cognitive modes as suggested by the decoupled representation of Chap. 4. In Example 1 for instance, the abbreviated representation EpN, extroverted perception Ep combined with intuitive perception N, suggests a 2-dimensional *extraverted intuition* combination Ne having components Ep and N. From context it is understood that the latter two variables are in the perception domain. Similarly in the same example, IjT in the judgment domain suggests a 2-dimensional *introverted thinking* combination Ti with components Ij and T.

As in Fig.4.4 for the attitudes, these relations can be plotted in coordinates that are radial, although not polar, so that all variables remain positive. In each domain the decoupled attitude is plotted vertically against the horizontal function coordinate, the mode axes being rotated 45% from the rectangular axes. These mode axes are symbolized naturally by %Se-%Ni and %Ne-%Si in the perception domain, and %Te-%Fi and %Fe-%Ti in the judgment domain.

Constructive geometric relations between the two coordinate systems will be developed in Sect. 5.3. They take the same mean sum "principal" and mean difference "subsidiary" form as for attitude decoupling. Just as for the

D. J. Wilde, *Jung's Personality Theory Quantified*,
DOI: 10.1007/978-0-85729-100-4_5, © Springer-Verlag London Limited 2011

attitudes, the subsidiary mode can be obtained by "unpacking" the larger variable, that is, subtracting the mean sum principal mode from it. In Sect. 5.4, graphical representations similar to those for attitude decoupling are developed. Brief descriptions of the eight modes follow in Sect. 5.5. All this is abstracted for further research in Sect. 5.7. Insightful inversion relations, dubbed "repacking" and "recoupling", are proven in Sect. 5.8. The concluding summary points the way to Chap. 6's discussion of Jung's dominant and auxiliary modes, together with the newer principal and subsidiary modes introduced in the present chapter.

5.2 Cognitive Mode Variables

Jung associated each cognitive mode with one decoupled attitude and one function. Table 5.1 shows the eight modes with their keywords. The mode formulas, each the simple average of the corresponding attitude and function, will be derived in Sect. 5.3. Note that opposite modes, Se and Ni for instance, are located diagonally opposite each other, as suggested by the arrows.

5.3 Data Combination

The relevant function and decoupled attitude data will now be combined to obtain the associated cognitive mode indicator scores. In Sect. 5.3.1 Example 1 will be compared to the ideal Extraverted iNtuitive "Ideator" in Table 5.2. All eight mode score formulas are displayed in Table 5.3. Following this is the Table 5.4 listing of the mode score formulas as first derived in Teamology (Wilde 2009).

Table 5.1 Cognitive modes with keywords

	Perception modes					Judgment Modes		
Se	*Ep*	*Ep*	*Ne*	-	*Te*	*Ej*	*Ej*	*Fe*
S	*Se* Experi-ment	*Ne* Ideation	*N*	-	*T*	*Te* Organiza-tion	*Fe* Commu-nity	*F*
S	*Si* Knowl-edge	*Ni* Imagina-tion	*N*	-	*T*	*Ti* Analysis	*Fi* Evalua-tion	*F*
Si	*Ip*	*Ip*	*Ni*	-	*Ti*	*Ij*	*Ij*	*Fi*

Table 5.2 Cognitive mode scores for the ideal ideator

Perception modes				Judgment modes			
Se	Ep	100%Ep	Ne	Te	Ej	Ej	Fe
S	0	100%Ne ideation	100%N	T	0	0	F
S	−100%Si	0	N	T	0	0	F
Si	Ip	Ip	Ni	Ti	Ij	Ij	Fi

Table 5.3 Cognitive mode score formulas

Perception modes					Judgment modes			
Se	*Ep*	*Ep*	*Ne*	-	*Te*	*Ej*	*Ej*	*Fe*
S	Se = (*Ep* + *S*)/2	*Ne* = (*Ep* + *N*)/2	*N*	-	*T*	*Te* = (*Ej* + *T*)/2	*Fe* = (*Ej* + *F*)/2	*F*
S	*Si* = (*Ip* + *S*)/2	*Ni* = (*Ip* + *N*)/2	*N*	-	*T*	*Ti* = (*Ij* + *T*)/2	*Fi* = (*Ij* + *F*)/2	*F*
Si	*Ip*	*Ip*	*Ni*	-	*Ti*	*Ij*	*Ij*	*Fi*

Table 5.4 Teamology cognitive mode score formulas

Perception modes				Judgment modes			
Se	Ep	Ep	Ne	Te	Ej	Ej	Fe
S	Se = E + P + 2S	Ne = E + P + 2N	N	T	Te = E + J + 2T	Fe = E + J + 2F	F
S	Si = I + J + 2S	Ni = I + J + 2 N	N	T	Ti = I + P + 2T	Fi = I + P + 2F	F
Si	Ip	Ip	Ni	Ti	Ij	Ij	Fi

5.3.1 The Platonic Ideal Ideator

In the style of the Platonic ideal reasoning of Sect. 4.3.1, the "ideal" ideator will have a maximum score 100% for extraverted intuition Ne, a minimum score −100% for its opposite mode introverted sensing Si, and zero for the other six modes. This must correspond to maximum scores of 100% for Ep and N.

To normalize as in Sect. 4.3.2, divide the attitude and function scores by 2 so that (100%Ep + 100%N)/2 = 100%Ne. For non-ideal values away from the maximum, simply interpolate linearly by replacing the 100% coefficients with the true percentages. This produces the general extraverted iNtuition interpolation equation Ne = (Ep + N)/2. Table 5.2 displays the eight cognitive mode scores for this ideal Ideator. The minimum definite % score would be (20%Ep + 20%N)/2 = 20%Ne. Smaller scores would be considered insignificant.

5.3.2 Mode Score Formulas

Since the same reasoning holds for the other seven modes, its details will be omitted here. The resulting general mode score formulas are shown in Table 5.3. The same form holds of course for all eight modes.

5.3.3 The Teamology Transformation

In slightly different form, all these relations were derived in Wilde 2009 and displayed in its Table 2.4 on p. 13. Table 5.4 here exhibits these "teamology" forms for comparison. The underscored variables in Table 5.4 are raw scores (not precentages) from the teamology questionnaire used in Chap. 2 here. Since the maximum values are 5 for the formula variable E etc., the mode score maxima must be $4 \times 5 = 20$ each. Hence the percentage mode scores of Table 5.3 can be obtained from the raw scores of Table 5.4 by applying the conversion factor $(100\%/20) = 5\%$. For instance Ne = Ne(5%), a relation whose derivation is left as exercise 5-7.

The conversion factor must be different of course for other score ranges. For MBTI raw scores, whose maxima are 30 instead of 5, the factor is $100\%/(30 \times 4) = 0.833\%$.

It may be of interest that the teamology formulas, forerunners of this entire book's project of quantifying Jung's theory, were developed as a simpler and faster alternative to the Type Dynamics way of estimating cognitive modes. The TD rules were considered too cumbersome to be applied to a class of forty or more students, so the teamology transformation was developed as an approximation for computing the modes on a spreadsheet for quick team construction. After a few years it became apparent that the teamology "approximation" often fit the personalities observed better than did the TD predictions that, as will be discussed in Chap. 7, are themselves only approximate at best. This insight generated the research underlying this book.

5.3.4 Example 1 Mode Scores

Table 5.5 gives the positive and zero mode scores for Example 1 (40%Ep, 100%N, 20%Ip, 20%T) evaluated from the formulas of Table 5.3. Negative scores are easily recovered as negatives of mode scores diagonally opposite. For reference though, the negative scores are shown.

Only for Ne and Ti are the attitude and function scores in the example both positive. For Ni the attitude score Ip is −40%, so $\text{Ni} = (\text{Ip} + \text{N})/2 =$

Table 5.5 Example 1b positive mode scores

Perception modes				Judgment modes			
Se	Ep	Ep	Ne	Te	Ej	Ej	Fe
S	−30%Se	70%Ne	N	T	0%Te	−20%Fe	F
S	−70%Si	30%Ni	N	T	20%Ti	0%Fi	F
Si	Ip	Ip	Ni	Ti	Ij	Ij	Fi

(−40% + 100%)/2 = 30%Ni as shown. The reader is advised to check the Te score since it also involves a negative score.

5.3.5 Mean Sum and Mean Difference

As it was for the decoupled attitudes in Chap. 4, one need not compute mode scores that might be negative. To avoid calculating negative modes, in each domain always work only with the positive scores. In the p-domain, these are the larger 100%N and the smaller 40%Ep, whereas in the j-domain they happen to be equal: 20%Ip and 20%T. In the p-domain the mean or average sum is thus (100%N + 40%Ep)/2 = 70%Ne, the mean or average difference being (100%N − 40%Ep)/2 = (100%N + (−40%)Ip)/2 = 30%Ni. The j-domain mean or average sum is (20%Ij + 20%T)/2 = 20Ti, while the mean or average difference vanishes in this case: %Te = (20%T + (−20%)Ej)/2 = 0.

5.3.6 Unpacking the Subsidiary Mode

If one considers the mean sum to be the "principal" mode score, a similar "subsidiary" mode score can be obtained by subtracting the principal score from the larger of the two variables involved, whether function or decoupled attitude. This "unpacking" has the same form as that derived in Sect. 4.5 for the attitudes, so its proof will not be repeated here.

When the function score exceeds that of the decoupled attitude, the subsidiary mode will have the same function (capital letter) as the principal mode, as in Example 1 with 40%Ne and 20%Ni. In such circumstances, the attitude subscripts must of course differ.

But when the decoupled attitude is larger, the subsidiary mode's function will be opposite to that of the principal mode, although the attitude subscripts will be the same. Thus if, as in Example 3 to come in Sect. 6.3, the variable values were reversed from those of Example 1 (100%Ep > 40%N instead of vice versa), the subsidiary mode would be Se instead of Ni, its score being the difference 30% between the larger 100%Ep and the mean 70%Ne. This is recapitulated in Sect. 6.3.

5.4 Graphic Representation

Now that the domains have been decoupled, they are each two-dimensional and amenable to well-known graphical representations. This brings visual understanding and computational insight that was not possible when the system was a coupled four-dimensional one.

5.4.1 *Elements Common to Both Domains*

In each domain the decoupled attitude can now be plotted against the domain function, the rectangular axes being "orthogonal", that is, at right angles. The mode axes, also at right angles, are rotated 45°, making them diagonal with respect to the rectangular axes. In teamological applications the diagonal modal system can be used for team formation (Wilde 2009, Chap. 3), whereas the rectangular coordinates are appropriate for team organization (Wilde 2010, Chap. 20).

An arcane technical comment: the coordinates are radial but not polar. For this reason the Cartesian quantities of angle and length have no meaning in these personality "spaces" where even the famous theorem of Pythagoras does not hold. Thus mathematicians would characterize personality geometry as "non-Euclidean". Mathematical novices need not be concerned about this, for the analysis to follow does not need Euclidean properties. Beware, however, of those who may naively speak of "distance" between personalities or "angles" between them, making improper analogies with Euclidean geometry and its associated vector analysis, a geometric model favored by engineers.

5.4.2 *Perception Domain*

The two-dimensional perception domain graph of Fig. 5.1 plots %Ep vertically against %N horizontally. The mode coordinates have %Se–%Ni versus %Ne–Si, the orthogonal axes being rotated 45° so that the corners of the two systems match Platonically as they did in Fig. 4.4 for attitude decoupling.

By ordinary Cartesian coordinate convention, the rectangular coordinates would have S–N horizontal and Ep-Ip vertical. The diagonal "modal"coordinates have Ne–Si from upper right to lower left, and Se-Ni diagonal from upper left to lower right. The region of slight scores (<20%) is shown as a light boundary square (Fig. 5.2).

The rectangular p-function and p-attitude points can be projected on to the diagonal coordinates to determine modal values. The mean sum is 70%Ne associated with Ep and N. The mean difference is 30%Ni associated with N and the smaller Ip. This smaller underscored mode is the subsidiary (p-)mode. A similar convention will hold in the j-domain.

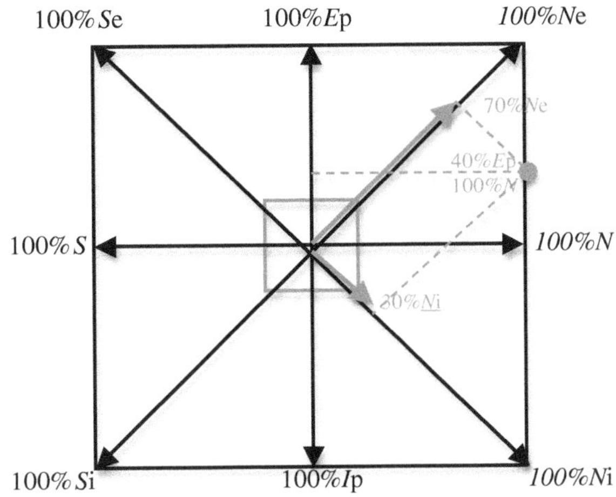

Fig. 5.1 Perception mode graph

Example 1a shown light.

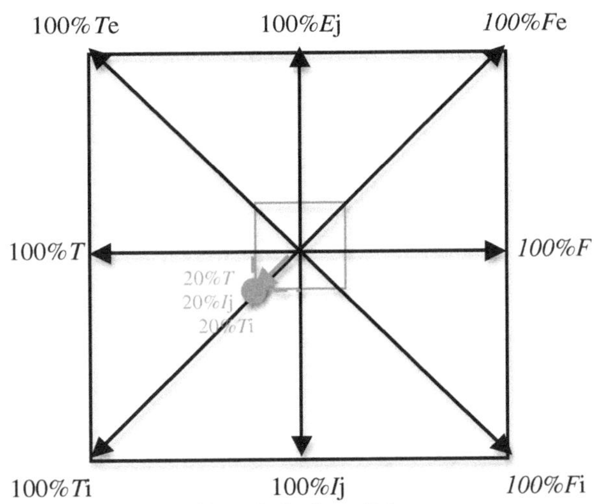

Fig. 5.2 Judgment mode graph

Example 1a shown light.

5.4.3 Judgment Domain

This subsection is just like that for the p-domain; only the variable names are different. The judgment domain also has two dimensions: %Ej versus %F, with rectangular orthogonal axes, and %Te–%Fi versus %Fe–Ti with diagonal orthogonal axes rotated 45°. As in the p-domain, these coordinates are radial but not polar, and the geometry is non-Euclidean.

In ordinary Cartesian coordinate terminology, the rectangular coordinates would have T-F horizontal and Ej-Ij vertical. The diagonal "modal"coordinates

Table 5.6 Cognitive mode descriptions (Wilde 2009, p. 26)

Se Experiment	Ne Ideation	Te Organization	Fe Community
Discovers new ideas and phenomena by direct experience	Rearranges known concepts into novel systems	Efficiently manages resources, decisive, imposes structure	Expressive, tactful builder of group morale
Si Knowledge	**Ni Imagination**	**Ti Analysis**	**Fi Evaluation**
Physically self-aware, values practice and technique	Prophetic, guided by inner fantasies and visions	Logically improves rational performance	Uses personal values to distinguish good from bad

have Fe–Ti from upper right to lower left, and Te-Fi diagonal from upper left to lower right. The region of slight scores (<20%) is shown as a light boundary square. Rectangular j-function and j-attitude points can be projected on to diagonal coordinates to determine modal values.

In Example 1 the mean sum is 20%Ti associated with Ij and T, and the mean difference is 0%Te/Fi associated with Ej and T/F.This underscored subsidiary j-mode vanishes in the example.

5.5 Cognitive Mode Descriptions

Table 5.6 gives short cognitive mode descriptions that have been found useful in teamological applications, the Ti Analysis description being due to Linda Berens. More complete descriptions have been offered by Haas and Hunziker (2006), Haas et al. (2001), and Thompson (1996).

5.6 Normalized Modal Score Representation

In the spirit of Sect. 4.9, two new ways of representing a numerical decoupled type come to mind. In rectangular coordinates each domain's decoupled attitude could be placed in front of the domain function, with the larger domain scores represented first. In this way Example 1b would be represented as: 40%Ep, 100%N; 20%Ij, 20%T.

In diagonal coordinates, the scores from the larger domain could be first, with both possible modes from each domain listed together. For each domain the smaller "subsidiary" domain would be underscored. In this way Example 1b would be represented as: 70%Ne, 30%Ni; 20%Ti, (0%Fi). Here the parentheses indicate a single-digit mode to be considered negligible. Notice the semi-colon (;) and extra space between the two domains.

Suppressing the numbers would produce symbolic modal representations. Example 1 would then read Ep, N; Ij, T in rectangular coordinates, and Ne, Ni; Ti, __ in diagonal ones. In this example the subsidiary j-mode, being zero, is an underscored blank.

As suggested in Sect. 3.5, **boldface** and lower case can be introduced to emphasize numerical differences in a semi-quantitative manner. This gives the rectangular Ep, N; Ij, T and diagonal **Ne, Ni**; Ti, __. Sect. 3.3's Example 1b (23%E, **93%N**, 13%t, **53%P**) is more interesting because its slight score for T generates the more varied representations 38%Ep, 93%N; 15%ij, 13%t and 66%Ne, 22%Ni, 14%ti, 1%te. Their semi-quantitative forms would be Ep, N; ij, t and **Ne**, Ni; ti, __.

5.7 Abstraction

Readers may have noticed that the presentations in Sect. 5.3 are identical in both domains except for the symbols. This of course is an invitation to combine the two presentations into a single unifying one usable more compactly in future applications such as that in the section to follow. To do this however will require going to the Greek alphabet for symbols, since the existing theory has used up most of the English letters that can be "mnemonic"—aids to the memory.

But the motivation here is not mathematical abstraction for its own sake. The purpose is to derive something useful for other psychological applications where two concepts are combined to form new ones. Jung's insight was to put together an attitude with a function to form something new of enhanced psychological value. His focus was on the larger cognitive mode generated, which was enough to make into an excellent personality theory. He did, however, overlook the smaller "subsidiary" mode that had to be generated in each domain, a mode not always negligible. This book calls attention to these subsidiary modes because they represent creative potential that could be lost if not recognized. In 2009 the University of Michigan's mathematical psychology professor Richard Gonzales pointed out to the author that other branches of psychology may suffer from similar oversights. The present subsection owes its inclusion to this possibility.

The presentation following suffers from the presence of two independent domains in the Jungian situation because this requires subscripts to distinguish between the domains. In a different application having only one domain, such subscripts would be unnecessary, making the analysis simpler.

5.7.1 New Terminology

This abstraction project needs new terminology that subsumes the pairs of concepts arising in the two domains. It can thus be useful for other psychological instruments and theories that combine variables to make new ones. Greek letters will be employed for any abstraction covering both domains.

Table 5.7 Greek letter terminology for abstraction

Variable Name	Variable Symbol	Greek letter Capital	Greek letter Lower case	Greek letter Name
Jung attitude pair	E & I	Γ	γ	Gamma
MB attitude pair	P & J	B	β	Beta
Decoupled attitude pairs	Ep & Ip	Δp	Δp	Delta
	Ej & Ij	Δj	δj	
Function pairs	S & N	Φp	ϕp	Phi, pronounced "fie"
	T & F	Φj	ϕj	
Cognitive mode pairs	Se &Ni	Mp	μp	Mu, pronounced "mew"
	Ne & Si	Mp	μp	
	Te & Fi	Mj	μj	
	Fe & Ti	Mj	μj	

Table 5.7 lists the capital and lower case Greek letters used, together with their somewhat mnemonic names in English letters and the variables they will represent here.

The subscripts p and j respectively represent the p- and j-domains in Jung's personology; usually they can be omitted during analysis. Upper case letters represent the larger, sometimes equal, number in a pair; lower case for the smaller number. This unfortunate source of confusion is needed to avoid laying out in detail the many cases of large and small that can occur. In Example 1a for instance, 20%E and 60%P will be represented respectively by γ and B. On the other hand, Example 2's 60%E and 20%P would be represented respectively by Γ and β.

5.7.2 Attitude Decoupling

Consider the attitude decoupling expressions of Eqs. (4.1 and 4.2). Abstract forms for them will have either Δ or δ equal to either (γ + B)/2 or (Γ + β)/2. Some examples may lessen the unavoidable confusion here.

In Example 1a the abstract form for the mean sum is Δp = (γ + B)/2 = (20%E + 60%P)/2 = 40%Ep. For the mean difference, δj = ($-\gamma$ + B)/2 = (($-$20%)I + 60%P)/2 = 20%Ij. For Example 2 the expressions are Δp = (Γ + β)/2 = (60%E + 20%P)/2 = 40%Ep again, together with the mean difference δj = (γ – B)/2 = (60%E + ($-$20%)P)/2 = 20%Ej, different this time from Example 1b.

5.7.3 Functions and Positive Cognitive Modes

Consider now the positive cognitive modes. The mean sum (primary and subsidiary numerical modes) formula is either μ = (δ + Φ)/2 or (Δ + ϕ)/2. Thus in example 1b, Δp = 40%Ep and Φp = 100%N, so the mean sum formulas give

μp = 70%Ne. The principal mode in the j-domain is μj = (Δj + Φj)/2 = (20%Ij + 20%T)/2 = 20%Ti.

The mean difference formulas for the subsidiary modes involve opposite variables, denoted by primes ($'$). These general formulas are (subscripts omitted for clarity) $\mu = (-\delta + \Phi)/2 = (\delta' + \Phi)/2$ or $(\Delta - \phi)/2 = (\Delta + \phi')/2$. In Example 1b, μp = (δp$'$ + Φp)/2 = ((−40%)Ip + 100%N)/2 = 30%Ni, and μj = (Δj$'$ − Φj$'$)/2 = (20%Ip − 20%F)/2 = 0%Fi (= 0%Te).

5.8 Inversion, Repacking and Recoupling

5.8.1 Preview

This section, which exploits the abstraction terminology developed in the section preceding, deals with *inversion*, the mathematical terminology for recovering the original MBTI scores from known cognitive mode scores. Inversion is immediately useful for checking quick or mental mode score calculations, a good habit for practitioners to develop. In the long run, inversion could be employed to evaluate new questionnaires, not currently existing, for determining the cognitive modes directly instead of from MBTI-style questionnaires.

Most important though is the insight the inversion equations provide. The original MBTI variables turn out to be simple sums or differences of the modes! Thus adding a pair of modes in the same domain gives either $\mu + \mu = (\delta + \Phi)/2 + (-\delta + \Phi)/2 = \Phi$ because the smaller attitudes δ cancel out in $(\delta + \Phi)/2 + (-\delta + \Phi)/2$, or $\mu + \mu = (\Delta + \phi)/2 + (\Delta - \phi)/2 = \Delta$ because the smaller functions ϕ cancel out in $(\Delta + \phi)/2 + (\Delta - \phi)/2$.

Similarly, subtracting a subsidiary mode from its principal mode gives either $\mu - \mu = (\delta + \Phi)/2 - (-\delta + \Phi)/2 = \delta$ because the larger functions cancel out in $(\delta + \Phi)/2 - (-\delta + \Phi)/2$, or $\mu - \mu = (\Delta + \phi)/2 - (\Delta - \phi)/2 = \phi$ because the larger attitudes Δ cancel out in $(\Delta + \phi)/2 - (\Delta - \phi)/2$. Subsequent recovery of the coupled Jung and MB attitudes from the decoupled attitudes thus involves only simple sums and differences of the decoupled attitudes, a process to be known as "recoupling".

5.8.2 Larger Variables

The larger functions Φ and attitudes Δ are obtained from either

$$\mu + \underline{\mu} = \Phi \quad \text{when } \Phi \geq \delta \tag{5.1Φ}$$

or

$$\mu + \underline{\mu} = \Delta \quad \text{when } \Delta \geq \phi \tag{5.1Δ}$$

In the p-domain of Example 1a, 100%N > 40%Ep, so μp + μp = 70%Nε + 30%Nι = Φp = 100%N. Thus in this case adding up the modes gives the original function, a relation very easy to remember. In the j-domain, where the j-function 20%T and j-attitude 20%Ij have equal scores, both versions of Eq. 5.1 hold, so 20%Ti + 0%Te = 20%T and 20%Ij. Inversion then amounts to "repacking" the mode scores to regain the original MBTI scores.

5.8.3 Smaller Variables

The smaller variables are obtained from

$$\mu - \underline{\mu} = \delta \quad \text{when } \Phi > \delta \tag{5.2δ}$$

or

$$\mu - \underline{\mu} = \phi \quad \text{when } \Delta > \phi \tag{5.2Φ}$$

Equality in the conditions is excluded to avoid ambiguity with Eqs. 5.1.

Equations 5.2 can be viewed as "unpacking" the larger principal mode μ to obtain the subsidiary mode μ plus the smaller original function ϕ or attitude δ. Thus the subsidiary mode acts as baggage arising from the inequality between the original variables. It is added to the smaller original variable to generate the principal cognitive mode μ. These results can be used to check the original numerical calculations.

In example 1, suppose only μp = 70%Ne, μj = 20%Ti, μp = 30%Ni and μj = 0%Fi have been given, it being desired to find the associated MBTI scores. Then the p-mode sums give the p-function N, in this case larger than the p-attitude Ep:μp + μp = 70%Ne + 30%Ni = Φp = 100%N.The j-mode sums give the j-attitude Ij, in this case also equal to the j-function T: μj + μj = 20%Ti + 0%Ti/Fi = Δj = 20%Ij(=20%T).

The p-mode differences give the smaller p-attitude Ep: μp − μp = 70%Ne − 30%Ni = δp = 40%Ep. The j-mode differences give the j-function T, here coincidentally equal to the j-attitude Ij:μj − μj = ϕj = 20%Ti − 0%Te/Fi = ϕ_j = 20%T (=20%Ij).

5.8.4 Attitude Recoupling

The preceding subsections worked entirely with decoupled attitudes. *Recoupling* is needed to recover the original Jung and M-B coupled attitudes from the decoupled ones. As it was for the modes, this involves sums or differences. For instance, the sum Ep +Ij = (E + P)/2 + (I + P)/2 = P because E and I cancel each other in the expression (E + P)/2 + (I + P)/2. The difference Ep −Ij = (E + P)/2 − (I + P)/2 = E because E = −I.

Table 5.8 Recoupling positive decoupled attitudes

	Ep	Ip
Ej	Ep + Ej = E Ep − Ej = P	Ip + Ej = J Ip − Ej = I
Ij	Ij + Ep = P Ij − Ep = I	Ip + Ij = I Ip − Ij = J

Table 5.8 enumerates the four possible cases of the two pairs of positive decoupled attitudes. A typical proof is given as exercise 5-8. In each case the coupled attitudes are given by the sum and the difference, with no division by 2.

In example 1a, 40%Ep and 20%Ij give 60%P and 20%E, which of course checks the original data, It is usually more convenient in practice to carry out the simple cancelations with the sum and difference than to look for or memorize Table 5.8. Another recoupling example (6) is given in Sect. 6.5.

5.9 Concluding Summary

This chapter, the heart of the book, gives amazingly simple two-term formulas for Jung's cognitive modes. They turn out to be average sums or differences of the well-known Jungian functions and the decoupled attitudes developed in Chap. 4. They are totally equivalent to the slightly more complicated three-term expressions derived for team construction theory, where the present quantification approach originated as a short cut for bypassing the qualitative approximations of Type Dynamics.

This striking simplicity deconstructs the original four-dimensional personality system into two completely separate problems, each with only two variables—a function and a decoupled attitude—easily but rigorously displayed on an ordinary graph. This plot exhibits two sets of coordinates, rectangular for the MB variables and diagonal for the cognitive modes. Results for one system are easily obtainable graphically by simple projection into the other. This is in fact how the transformation equations are derived.

Going backwards from results to data, "inverting the transformation"that is, shows that the modes may be considered "unpackings" of the original variables into either various attitudes of a function, or various functions with the same attitude. Since the theory is the same for both domains, the abstraction for other psychological applications is straightforward. The ideas here should therefore apply whenever new variables are formed by combining earlier ones. This of course is what Jung did when he conceived the cognitive modes.

Much of this chapter echoed Chap. 4: domain independence, Platonic reasoning, mean sums and mean differences, unpacking, rectangular and diagonal coordinate systems, and the graphics. New features were the development of an abstract theory for new applications and inversion for numerical checking.

This chapter lays the foundation for Chap. 6's discussion of Jung's dominant and auxiliary modes, as well as the newer concept of subsidiary modes.

5.10 Exercises

5-1 Calculate the positive normalized cognitive mode scores from your results on the teamology questionnaire.
5-2 If you have MBTI scores, calculate your mode scores from them.
5-3 Do the same for Example 2: 60%E, 20%P, 100%N, 20%T.
5-4 Calculate the mode scores for Example 3 to come in Sect. 6.3: 100%E, 100%P, 40%N, 20%T.
5-5 Calculate the mode scores for Example 4 to come in Sect. 6.5: 100%I, 100%S, 100%T, 100%J.
5-6 Calculate the mode scores for Example 5 to come in Sect. 6.6: 100%E, 60%J, 60%S, 20%F.
5-7 Carry out the substitutions and algebraic manipulations needed to prove that for the teamology transformation, Ne = Ne(5%) as asserted in Sect. 5.3.3.
5-8 Prove the result in Table 5.8 for positive decoupled attitudes Ip and Ej. Check your result for the case 50%Ip and 20%Ej.

Chapter 6
Dominant, Auxiliary, Principal, and Subsidiary Modes

They kept asking for more.
—last words of the matador Manolete

6.1 Introduction

The preceding chapter showed that quantifiying Jung's personality theory inevitably leads to two modes in each domain—four in all. Jung himself considered only the principal one in each domain, giving the name "dominant" to the more prominent and "auxiliary" to the other. Quantitative theory identifies a new "subsidiary" mode in each domain, both absent from Jung's entirely qualitative formulation. In another context, Jung regarded them as "positive shadows" (Jacobi 1942, p. 112), an interpretation discussed further in Sect. 8.3.

The chapter has many numerical examples displayed in tables designed to show how the functions and attitudes unpack as they combine to form cognitive modes. A surprise is that any (Platonic) ideal Myers–Briggs four-letter type must have a subsidiary mode in the auxiliary domain. An especially interesting situation occurs when quantitative methods enter the picture. Of the two variables combining to form a cognitive mode, only the larger one can be unpacked to yield the subsidiary mode. The smaller variable score happens to be the difference between the principal and subsidiary mode scores, a useful checking relation and one that will lead to a practical rounding strategy. A college seminar example will have the principal and subsidiary modes pack in one domain into a function and in the other into an attitude. The new table form will show all this quite clearly, and a few words are devoted to the minor details, heretofore overlooked, of slightness and rounding.

Since for Jung the dominant mode was the most prominent, distinguishing the dominant one from the auxiliary is difficult when two modes have similar scores. This dilemma generates a grey zone that may not be resolvable by unaided quantitative analysis. If one affixes the word "dominant" not only to the largest score cognitive mode, but also to the function and the decoupled attitude with the largest scores, the three items may not always correspond. From the quantitative point of view then, the words "dominant" and "auxiliary" may need to be replaced with "principal" and "secondary" to avoid ambiguity.

D. J. Wilde, *Jung's Personality Theory Quantified*,
DOI: 10.1007/978-0-85729-100-4_6,

The *MBTI* type table unfortunately overlooks the human potential hiding in these subsidiary modes. Teamology (Wilde 2009), which does recognize them, has in fact successfully used them to construct better teams and improve the performance of existing ones. A similar approach should indeed prove valuable for understanding any interpersonal situation interesting to counselors and 8-function (-attitude) personologists. On the other hand, the chapter offers an anecdote suggesting caution in team situations when assigning subsidiary modes.

6.2 Dominant and Auxiliary Modes

In Example 1 the greater attitude score 40%*E*p (>20%*I*j) and the greater function score 100%*N* (>20%*T*) are both in the same (perception) domain. In this case the corresponding mode will have largest mode score, so 70%*N*e would be the dominant Jungian type. Intuition would be the dominant function, not just because it is associated with the dominant mode, but also because it has the larger function score. The smaller function score 20%*T* then identifies Thinking as the auxiliary function in this case.

Jung stated that dominant and auxiliary modes must be in different domains. This makes Thinking the auxiliary mode in Example 1 because it is the only significant mode in the opposite (*j*-) domain, even though there is another mode 30%*N*i in the *p*-domain having a larger score. Jung didn't consider the attitudes in his dominance formulation.

The quantification of Jung's personality theory, together with the decoupling of the attitudes, makes it possible that the greater attitude (marked #) and greater function (marked *) can be in different domains. When this happens, formal dominance is ambiguous, so the quantitatively precise terms "principal" and "subsidiary" will be used in place of "dominant" and "auxiliary". This situation occurs in Example 5 (Sect. 6.6) and is discussed in Sects. 7.3.1 and 7.3.2. Incidentally, since Teamology works entirely with mode scores to construct teams, it does not employ or need the concept of dominance.

6.3 Subsidiary Modes

The rules of Type Dynamics only identify "dominant" and "auxiliary" modes. Although Sect. 7.3 shows that this identification is not always correct, that is not the point here. It is rather that *TD*'s qualitative approach cannot detect subsidiary modes such as Example 1's 30%*N*i. In principle there is such a subsidiary mode in each domain, two in all, although the smaller one might be negligible as in Example 1's *j*-domain *T*e–*F*i pair.

Subsidiary modes will usually be underscored. Significant subsidiary modes occur whenever attitude and function differ significantly. Thus in Example 1, 100%*N* > 40%*E*p gives 70%*N*e and 30%<u>*N*i</u>, an unpacking of the larger variable *N*.

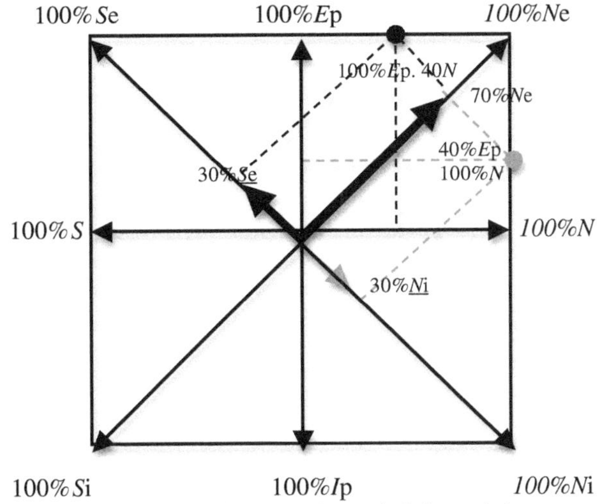

Fig. 6.1 Subsidiary modes for Examples 1 and 3

Example 1 shown light. Example 3 shown dark.

The resulting subsidiary mode, which has the same function but opposite attitude, is shown light in Fig. 6.1. This subsidiary mode 30%Ni (imagination) is an example of what Jung calls a "positive shadow" (Jacobi, p. 112). "Shadow" is the Jungian term for

> the contents (sometimes positive) of our psyche that have been rejected and repressed or less lived in our conscious existence…

Chapter 8 will discuss shadow aspects more fully, especially as they relate to subsidiary modes.

A new Example 3 may aid the understanding of subsidiary modes. In it the scores for Example 1's *p*-function and *p*-attitude have been interchanged so that 100%*E*p > 40%*N* gives 70%*N*e and the subsidiary 30%Se, contrasting with the 30%Ni subsidiary in Example 1. Notice that the larger attitude now yields a subsidiary mode having the same extraverted attitude but a function *S* opposite to that of the primary mode, which is unchanged from that of Example 1. Here Se (experiment) is a positive shadow, shown dark in Fig. 6.1.

Subsidiary modes have been quite useful in Teamology and are not to be overlooked. They can occur as auxiliary subsidiaries as well as dominant subsidiaries and having two subsidiary modes is not uncommon.

6.4 Slightness and Rounding

6.4.1 Slight Modes

To understand the effect of slight modes, consider Example 1b, even though the numbers are no longer round: 23%*E*, 93%*N*, 13%*T*, 53%*P*. Since *P* > *E*, the

Table 6.1 Attitude table, Example 1b

	P 53% Flexibility	*J* Structure
E 23% Extraversion	***Ep*** **38%# Exploration**	***Ej*** **Control**
I Introversion	***Ij*** **15% Appraisal**	***Ip*** **Focus**

Coupled attitudes are shown in normal. Decoupled attitudes are shown in bold

Table 6.2 Mode table, Example 1b

Perception			Judgment		
	S Sensing (Facts)	*N* 93%* intuition (Concepts)	–	*T* 13% Thinking (Things)	*F* Feeling (People)
Ep 38%# Exploration	***Se*** **Experiment**	***Ne*** **65%#* Ideation**	– *Ej* Control	***Te*** **Organization**	***Fe*** **Community**
Ip Focus	***Si*** **Knowledge**	$\underline{\mathit{Ni}}$ **28% Imagination**	– *Ij* 15% Appraisal	***Ti*** **14% Analysis**	***Fi*** **1% Evaluation**

Decoupled attitudes & functions are shown in normal. Cognitive modes are shown in bold

decoupled attitudes are $\%Ep = (23\%E + 53\%P)/2 = 38\%Ep\#$, and $\%Ij = (-23\%E + 53\%P)/2 = 15\%Ij$. Attitude Table 6.1 displays these results.

For the *p*-modes, the intuition function *N* is marked with * because it exceeds the *j*-function *T*. Since $N^* > Ep\#$, the appropriate formulas are $\%Ne = (38\%Ep\# + 93\%N^*)/2 = 66\%Ne\#^*$ (rounded down), and $\underline{\%Ni} = (-38\%Ep + 93\%N)/2 = 27\%\underline{Ni}$ (rounded up). Being marked with both # and *, extraverted iNtuition is clearly the dominant mode. There are two intuition modes adding up to the original *N* score, as will be explained at the end of the section.

In the *j*-domain, $Ij > T$, so the appropriate formulas are $\%Ti = (15\%Ij + 13\%T)/2 = 14\%Ti$, and $\underline{\%Fi} = (15\%Ij - 13\%T)/2 = 1\%\underline{Fi}$. There happen to be two introverted judgment modes, although $\underline{Fi}$ is negligible. Their sum is the *j*-attitude score, as will soon be explained.

Mode Table 6.2 summarizes and organizes these calculations in a way intended to display clearly the personality represented. Even though introverted thinking *Ti* is slight, it must be auxiliary because it is the only mode not in the dominant domain. Even though listed, the 1%*Fi* is of course negligible.

The literal representations exhibiting slightness and *p*-dominance are *EpN*; *ijt* in rectangular coordinates. *Ne* $\underline{Ni}$; *ti* does this in diagonal coordinates.

6.4.2 Rounding

Sharp-eyed readers will notice that strictly speaking the *N* modes are half a percent off from the formulas of Table 5.3. In fact, *Ne* was rounded down while $\underline{Ni}$ was

Table 6.3 Attitude table, Example 4

	P Flexibility	*J* 100% Structure
E Extraversion	***E*p Exploration**	***E*j 0% Control**
I 100% Introversion	***I*j Appraisal**	***I*p 100%# Focus**

Coupled attitudes are shown in normal. Decoupled attitudes are shown in bold

rounded up. This admittedly arbitrary rounding strategy has the advantage of making the mode percentages add up to the function score, a relation that follows from carrying out the algebra for $N\text{e} + \underline{N\text{i}} = (E\text{p} + N)/2 + (I\text{p} + N)/2 = N$, in which the extraverted and introverted attitudes cancel, leaving only the function remaining.

A similar but slightly different situation arises when the decoupled attitude is larger than the function, as in the *j*-domain of Example 1b. Adding the scores for modes with the same attitude will in this case produce the original decoupled attitude score because the functions cancel. Thus in the judgment domain of Example 1b, $T\text{i} + F\text{i} = (I\text{j} + T)/2 + (I\text{j} + F)/2 = I\text{j}$. Rounding doesn't happen to be an issue here, but if it were, the suggestion would be to round the principal mode down and the subsidiary mode up.

It follows that in general a good mode computation strategy would be to use Table 5.3 only to calculate the principal modes in each domain, always rounding down. Subsidiary modes then would be found by subtracting the principal score from that of the larger variable, be it function or attitude.

6.5 A Platonic Ideal *MBTI* Type (Example 4)

With the help of another "Mode Table", the generic name for tables like Table 6.2, the Platonic Ideal for the Myers–Briggs types will now be studied. The particular type chosen as an example will be the *ISTJ* type, its Platonic scores being the maxima 100%*I*, 100%*S*, 100%*T* and 100%*J*. Attitude Table 6.3 shows that its decoupled attitudes are 100%*I*p and, interestingly, 0%*E*j, which give the verbal interpretation of ideal introverted perception and NO extraverted judgment. Actually though, "no extraverted judgment" really says that the *j*-function *T* is as likely to be introverted as extraverted, as will be shown.

Such a Platonic ideal *ISTJ* person actually exists, and his interaction with the Example 1 person will be discussed after Mode Table 6.4 is presented. The subject is a well-known type educator and author.

The tie between 50%*T*e and 50%*T*i requires a decision as to which is auxiliary for purposes of the archetypal analysis of Chap. 10. Here the Type Dynamics attitude balance assumption is as good as any, so take the extraverted mode *T*e

Table 6.4 Platonic *ISTJ* mode table

Perception			Judgment		
	S	*N* –		*T*	*F*
	100%* Sensing (Facts)			100% Thinking (Things)	
*E*p	***S*e**	***N*e** –	*E*j	***T*e**	***F*e**
			Control	**50% Organization**	
*I*p	***S*i**	***N*i** –	*I*j	$\underline{Ti}$	***F*i**
100%# Focus	**100%#* Knowledge**		Appraisal	**50% Analysis**	

Decoupled attitudes & functions are shown in normal. Cognitive modes are shown in bold

(Organization) as auxiliary and *j*-principal, leaving the introverted $\underline{Ti}$ (Analysis) mode to be considered *j*-subsidiary, a *positive shadow*.

This *T*i mode became important in an e-mail exchange with the Example 1 person, for whom this mode is auxiliary. The mode was first experienced by the *ISTJ* as an archetypal *negative* shadow, his earliest reaction being automatic rejection of the new ideas embodied in quantification. But soon his positive analytic quality emerged that connected easily with the analytic component of the Example 1 personality. This subsequent unusually productive exchange between an *ISTJ* and an *ENTP* will be discussed further in Archetypes Chap. 8.

6.6 Ambiguous Dominance (Example 5)

6.6.1 Quadmodal Example 5

Example 5 is a case having four significant modes, three of them extraverted. It describes a Stanford sophomore from the Teamology seminar there who carried out duties successfully in all four modes. The second team on which he appeared earned the 2009 "Best-of-Seminar" award on their first and only project near the end of the quarter. His scores from a second test near the end of the quarter were: 100%*E*, 60%*J*, 60%*S*, 20%*F*, (*ESFJ*). Attitude Table 6.5 shows that the decoupled attitudes, both extraverted, are *E*j# = 80% and *E*p = 20%.

The dominance is ambiguous here because the larger decoupled attitude score 80%*E*j# (>20%*E*p) is in the *j*-domain, whereas the larger function score 60%*S** (>20%*F*) is in the *p*-domain. For this reason the words "principal" and "subsidiary", applied to both domains as appropriate, are more descriptive here than "dominant" and "auxiliary" alone. Table 6.6 is the resulting Mode table.

This situation is written abstractly as $\Delta j > \delta p$ but $\Phi p > \phi j$. The decoupled attitudes are $\delta p = 20\%Ep$ and $\Delta j = 80\%Ej\#$. The perception modes are $\mu p = 40\%Se*$ and $\underline{\mu p} = 20\%\underline{Si}$, making the *p*-domain bimodal. The judgment modes are $\mu j = 50\%Fe\#$ and $\underline{\mu p} = 30\%\underline{Te}$, so the *j*-domain is also bimodal.

Consequently the subject is a quad-modal double extravert with the following mode identifications: $\mu j = Fe\#$ *j*-principal, $\underline{\mu j} = \underline{Te}$ *j*-subsidiary; $\mu p = Se*$

Table 6.5 Attitude table, Example 5

	P	*J* 60% Structure
E 100% Extraversion	***Ep* 20% Exploration**	***Ej* 80%# Control**
I	***Ij***	***Ip***

Coupled attitudes are shown in normal. Decoupled attitudes are shown in bold

Table 6.6 Example 5 mode table

Perception			Judgment		
	S 60%* Sensing (Facts)	*N*	–	*T*	*F* 20% Feeling (People)
Ep 20% Exploration	***Se* 40%* Experiment**	***Ne***	– *Ej#* 80%# Control	***Te* 30% Organization**	***Fe* 50%# Community**
Ip	***Si* 20% Knowledge**	*Ni*	– *Ij*	*Ti*	*Fi*

Decoupled attitudes & functions are shown in normal. Cognitive modes are shown in bold

p-principal, $\underline{\mu p}$ = $\underline{Si}$ *p*-subsidiary. The literal description in diagonal coordinates is *Fe#*, $\underline{Te}$; *Se**, $\underline{Si}$, with all modes significant.

6.6.2 An Uncharted Personality?

Since this personality does not quite fit into Jung's simpler model organized around a clearly dominant mode, it may be useful to discuss the subject's behavior on his first team, which carried out four hands-on construction assignments. Although all three teams in the Teamology seminar performed well, the subject's team seemed to lack coherence in the sense that it did not integrate the ideas of its four members very well into single plans of action.

This is important here because the subject's extraverted feeling *F*e Community mode was expected to catalyze the team's integration, the very thing that was lacking. In retrospect, the team may have made an organizational mistake in asking the subject to cover, in addition to the obvious "Diplomat" *F*e role, the neighboring "Coordinator" $\underline{Te}$ role, both roles for which the subject exhibited interest.

With 20–20 hindsight one can speculate that it would have been better to have had the subject concentrate on pulling the team together, leaving the coordination to a slightly quieter team-mate favoring his Thinking function (Exercise 6-3). As mentioned earlier, when the subject was on a later team on which the Coordinator role was filled by someone else (Exercise 6-4), the new organization turned out the best project of the quarter.

This digression is not intended to illuminate Jung's existing theory. It is instead put forth as a comment on how the increased complexity brought about by quantification may expose personality subtleties overlooked by Jung's simpler qualitative formulation. It thus opens up new research possibilities that would combine rigorous computation with careful observation in practical interpersonal situations.

6.7 Recoupling

To invert the original problem, suppose that the four mode scores were the only information available, say by an as yet uninvented questionnaire. The inverse problem then would be to find the associated *MBTI* scores. Scores for the functions and decoupled attitudes are easily determined from a table with the same form as Table 6.6 but with only the mode scores filled in, as in Table 6.7.

In either domain, each mode pair will occupy either a full row, as in the *j*-domain in Example 5, or a full column as in the *p*-domain. Adding up row entries (30%*T*e and 50%*Fe*) will produce the decoupled attitude score (80%*E*j#) as in Table 6.6, The smaller *j*-variable is the row difference (50 – 30)% = 20%, entered into the *j*-function cell *F* next to the *j*-principal cell 50%*F*e. Similarly, adding column entries 40%*S*e and 20%*S*i gives the function score 60%*S**. The smaller *p*-variable is the column difference (40 – 20)% = 20%, entered into the *p*-attitude cell next to the *p*-principal cell 40%*S*e.

To regain the original coupled attitudes *E–I* and *J–P*, the decoupled attitudes 20%*E*p and 80%*E*j must be recoupled as discussed in Sect. 5.8.4. Rather than refer to Table 5.7 there, merely redraw Attitude Table 6.5 with the decoupled attitudes filled into two adjacent squares as in Table 6.8.

If as in the example, they are in the same row, their sum (20%*E*p + 80%*E*j) will produce the Jungian attitude score for that row (100%*E*), verifying Table 6.5. The difference between the two decoupled scores (80%*E*j – 20%*E*p) is the Myers–Briggs attitude score in the primary attitude column marked by the # (60%*J*).

On the other hand, if the decoupled attitudes scores are in the same column the score sum will yield the *M–B* attitude score, and the difference will be the Jung

Table 6.7 Example 5 mode table with unknown functions and attitudes

	Perception			Judgment		
	S	*N*	–		*T*	*F*
	?					?
*E*p	**Se**	***N*e**	–	*E*j	**Te**	*F*e
?	**40%* Experiment**			?	**30% Organization**	**50%# Community**
*I*p	**Si**	***N*i**	–	*I*j	***T*i**	***F*i**
	20% Knowledge					

Decoupled attitudes & functions are shown in normal. Cognitive modes are shown in bold

Table 6.8 Attitude Table, Example 5 with unknown coupled attitudes

	P	*J*
		?
E	*E*p	*E*j
?	**20% Exploration**	**80%# Control**
I	*I*j	*I*p

Coupled attitudes are shown in normal. Decoupled attitudes are shown in bold

attitude score for the row with the #. This happens in Example 1b, for which the attitude table is Table 6.1 in Sect. 6.4.1.

6.8 Concluding Summary

This chapter goes beyond the subtle simplifications of Jung's qualitative theory to expand, perhaps double, the power of the *MBTI* to describe human personalities. The major expansion comes from recognizing the new subsidiary modes, one in each domain, that display human potentials that Type Dynamics cannot detect. Jung saw them as positive shadows supplementing the ego. School counselors in particular are advised to beware of losing student potential because of qualitative oversimplification of the *MBTI* scores.

Quantitative analysis also shows that even the concept of dominance can be ambiguous. For greater precision it is recommended that the terms "principal" and "subsidiary" be substituted for Jung's "dominant" and "auxiliary" whenever a person's dominance relations are unclear.

The chapter also presented a compact tabular representation of the modes that clearly displays the packing and unpacking phenomena. Minor but not unimportant questions concerning slight variables and numerical rounding were cleared up.

This completes the development of a new quantitative expansion of Jung's personality theory. The next chapter will derive its relation to the more limited Type Dynamics categorical approach, showing how to detect the many situations in which *TD* simplifications are in fact appropriate and economical. This will lay the foundation for the Chap. 8 an update of Beebe's archetype theory, which has unfortunately been called into question lately because of its earlier connection with unmodified Type Dynamics.

6.9 Exercises

6-1 Display your teamology questionnaire results in attitude and mode tables like Tables 6.1 and 6.2.

6-2 If you have *MBTI* scores, display your results in attitude and mode tables like Tables 6.1 and 6.2. Point out any rounding necessary.

6-3 On his first team, the Example 5 student in Sect. 6.6 had a team-mate with questionnaire scores *I*1, *J*4, *S*5, *T*5. Construct the latter's attitude and mode tables.

6-4 On his second team, the Example 5 student in Sect. 6.6 had a team-mate with questionnaire scores *E*3, *J*0, *S*4, *T*5. Construct the latter's attitude and mode tables.

Chapter 7
Partially Fixing Type Dynamics

He who makes a mistake, and doesn't correct it, makes two mistakes.
—Kong Fuzi (Confucius, 551-479 BCE)

7.1 Introduction

This chapter is dedicated to the correct, or one might say "corrected", use of the traditional Type Dynamics (*TD*) approach to personality typing. It will show, not only how to tell when *TD* will work, but also what to do when its uncorrected performance is inadequate.

The chapter is written to be understood even by those who may not have mastered the preceding chapters, provided they can accept those results without needing to understand their origins and proofs. Wherever possible the discussion is expressed in qualitative terms involving *MBTI* letter categories rather than numerical scores. Only two conditions involve simple numerical comparisons of scores, only rarely the same, and for different variables. By confining itself to different variables, this practice deliberately avoids attributing measurement properties to the *MBTI*.

Although *TD* never intended to recognize modes other than Jung's usual dominant and auxiliary ones, subsidiary modes are in fact easily detectable when the attitudes are decoupled. This extends the scope of *TD*, greatly expanding its usefulness for career and interpersonal counseling.

The conditions in which *TD* is not directly applicable are determined from the quantitative analysis of the earlier chapters. When inapplicability occurs, for perhaps a fifth of the clientele, simple calculations can generate true conclusions. These conditions are interesting in themselves because they increase the understanding of personality theory. One condition is that *TD* can work only if the Myers–Briggs attitude *P–J* score is at least as large as the Jungian *E–I* score. The other is that the score of the function in the domain of the larger decoupled attitude be at least as large as that of the other function.

Isobel Briggs Myers and her mother Katherine Cook Briggs rightly deserve plaudits for developing and launching the *MBTI* in the face of initial opposition from academic psychologists and even Jungians. Since then hundreds of millions of people all over the world have used the instrument, many of them thereby

D. J. Wilde, *Jung's Personality Theory Quantified*,
DOI: 10.1007/978-0-85729-100-4_7,

discovering valuable but previously unrecognized parts of their personality. A sizable profession of certified *MBTI* practitioners has grown up to administer the instrument, accompanied by publishers and training schools supported by royalties distributed through the Myers Foundation, who sees itself as a dedicated guardian of the Myers reputation and legacy.

This community of counselors and educators was badly shaken by the recent revelations of experimental psychologists Reynierse and Harker (2008), who found that *TD* mappings to cognitive modes simply did not fit the 700 or so data items they examined. Reynierse (2009) also identified basic categorical errors in *TD*'s logic in his provocatively titled article "The Case Against Type Dynamics". Many certification educators and publishers resisted his remarks, considering them an attack on the traditional method of analysis they had used for almost half a century.

Coincidentally, or "synchonistically" as Jung might say, that same year the book *Teamology* appeared as a guide to constructing and organizing design teams, indirectly aided by the *MBTI*. Forerunner of the present book, *Teamology* avoided the errors of *TD* by analyzing the data quantitatively instead of qualitatively. Soon it became clear that one could deal with the *TD* problems exposed by Reynierse and Harker simply by quantifying the analysis as developed in the present book's earlier chapters. Consequently this book was written, aimed at the *MBTI* community unlikely to be drawn to *Teamology*'s focus on teams.

This chapter does more than explain and correct the *TD* errors exposed by Reynierse and Harker. It also builds a modified procedure that not only recognizes when *TD* gets things right, but also knows how to correct the situation when errors creep in. Clarifying the logic behind *TD*-style operations generates a couple of simple conditions that most of the time preserve the *TD* procedure while keeping it out of trouble in anomalous cases. This should enable traditional type dynamicists to adapt to the more general circumstances with little effort.

Newcomers to *MBTI* work may of course prefer to dispense with *TD* entirely so as to concentrate on the more straightforward theory of the preceding chapters. But certified counselors can also in good conscience use this chapter's approach to take advantage of the increased power of the *MBTI* made available by slight modification of the *TD* approach with which they have so much experience.

This gives type educators several choices on how to incorporate quantification into their certification courses. Previously certified practitioners can be offered "post-graduate" courses that follow either the full quantification theory of the earlier chapters or else the *TD* modification approach of the present chapter. Eventually newcomers to the field could learn everything in one revised course.

After a brief section honoring the *MBTI* pioneers, acknowledging the foresight and initiative that led to the simplifying approximations needed to get things started, the chapter discusses the subtleties of logic, notably categorical reasoning, that generated the errors corrected here. Then a pair of examples, one fitting the *TD* model and the other not, are used repeatedly to show the limitations of the various rules of Type Dynamics. This leads to resolution of a misreading of Jung that appears to cause uncomfortable contradictions between the qualitative and quantitative theories. Implications for the famous Type Table are then discussed together with a hypothetical suggestion for its expansion. Personality variety previously hidden within each *MB* type is then detected by quantitative analysis. Finally the modes opposite the dominant and auxiliary are discussed in preparation for Chap. 8, in which Beebe's Archetype theory is reviewed as a clinical rather than a numerical key to shadow modes both negative and positive. The chapter ends by recognizing Reynierse and Harker, not only for their *preference multi-dimensionality* approach to typology proposed as a substitute for Type Dynamics, but also for their courageous distress signal that *TD* needed immediate repair.

7.2 Homage to Isabel Myers, *MBTI* Pioneer

New endeavors usually are led by pioneers who bravely push past obstacles into the dangerous future, where they must make approximate estimates (guesses) to get started. Later, to continue progress, their followers must find and correct the earlier mistakes. Correcting such errors means no disrespect to those whose vision and enterprise were essential to get things started.

Isobel Myers deserves special recognition for her initiative in developing and testing the *MBTI*, her willingness to guess and approximate, her literary skill in writing type descriptions and manuals, her enterprise in promoting the *MBTI*, her making the *MBTI* simple and easy to use, and her development of talented supporters to carry on the work. The analysis hereof her early work is intended to be fully constructive, offered to bring the underrated *MBTI* the full respect it deserves, for it is twice as powerful as even Myers believed. This chapter has been written entirely in a positive spirit of advancing the practice of *MBTI* personology.

7.3 Categorical, Comparative and Computational Reasoning

The traditional qualitative approach to analyzing the *MBTI* instrument uses only the four pairs of type letters, but never the data, the numerical scores associated with the letters. Logicians call this approach "categorical" because it works entirely with discrete categories to which a subject either belongs or doesn't belong. Partial membership is not acknowledged.

Conventionally the *M–B* data are used only once, for indicating membership in the various type categories. For some time there has been a controversy among personality theorists about whether or not a category variable's score measures degree of membership in the category. Traditional *MBTI* theorists oppose the measurement idea to the point of imposing a taboo in their certification programs against using score data in any way. This book contends, however, that the scores can and should be used for many things other than trait measurement.

Moreover, comparing scores between a single person's different variables is very different from comparing same variable scores for two different people. Just as the signs of the primitive scores determine *MBTI* category membership, simple combinations of scores can establish membership in decoupled attitude and Jungian cognitive mode categories useful for counseling and 8-function (-attitude) personology. This type of reasoning, which compares numbers representing different categories to determine some categorical property, will be known here as *comparative* reasoning. A little comparative reasoning will neatly plug the holes in the purely categorical *TD* scheme.

A further reasoning step to fill *TD* gaps concerns actual calculation of numbers associated with the various categories. Such *computational* reasoning, which fills the preceding chapters, is here needed only to extend the reach of Type Dynamics so that it can identify more modes than the usual two, a purpose for which it was not originally designed or intended.

The next subsection deals with and corrects the unfortunate confusion, familiar to experienced *MBTI* practitioners, introduced by the pioneers when they named their new attitudes. They unintentionally committed what logicians call "categorical errors" when they gave the same labels to their new attitudes as they gave to Jung's two domains of cognitive modes.

7.3.1 Categorical Error

Early on, the term "perception" was given not only to the domain of perception modes, which Jung by the way had previously called "irrational", but also to the new attitude symbolized by "*P*". Similarly, the term "judgment" was given not only to the domain of judgment modes (Jung labeled them "rational") but also to the new "*J*" attitude. This categorical error (committed twice) of giving the same name to two different categories was pointed out by Reynierse (2009). It has been confusing practitioners throughout the history of the *MBTI*. The book *Teamology* circumvented this problem by renaming the domains: "Information Collection" for the *p*-domain and "Decision-making" for the *j*-domain. In retrospect this opportunistic strategy seems a bit evasive, avoiding the problem rather than confronting it.

Chapter 4 gave these attitudes artificial meaningless names "pea" and "jay" for temporary use during the development of the new decoupled attitudes. The new attitude pairs were given the new names Extraverted and Introverted Perception and Extraverted and Introverted Judgment, symbolized respectively by *E*p–*I*p and *E*j–*I*j, Thereafter the new decoupled attitude pair labels were substituted for the original coupled attitude pair labels *E–I* and *P–J*. This was guided by Jung's implied but numerically unrealized original formulation discussed in Chap. 4. These new decoupled attitudes are of course free from the categorical naming error of the old coupled attitudes, finally removing the original confusion.

Two parallel examples will be used throughout the theoretical development to follow. A new Example 6 (20%*I*, 20%*N*, 60%*T* 60%*P*) will be valid for analysis by the usual *TD* rules without modification. The second, Example 5 (100%*E*, 60%*S*, 20%*F*, 60%*J*) from Sect.6.6, will need modifications to the usual *TD* rules to get correct results. The latter example is intended to show where the problems are with *TD* so that experienced practitioners will see where and how to fix things.

7.3.2 Principal Decoupled Attitude

To begin, let *D* represent the *principal* decoupled attitude, identified quantitatively by the higher score, although the score will usually not be computed here. Denote the *E–I* attitude generically as *C* (for Carl), and the *P–J* attitude as *B* (for Briggs). This extra notation is introduced to cut down the number of cases needed to cover the subject. It will be shown that *C* and *B* can determine the principal decoupled attitude *D* entirely categorically, without any comparisons or computations.

First consider any extraverted type, for which $C = E$. Then typologists agree that in addition $B = P$ implies $D =$ *E*p (Extraverted Perception), and $B = J$ implies $D =$ *E*j (Extraverted Judgment) as in Table 7.1. Notice that coincidentally the subscripts match the *MB* letter for "Extraverts". This does not happen for "Introverts" however, for which $C = I$. This time $B = P$ implies $D =$ *I*j and $B = J$ implies $D =$ *E*p as in Table 7.1. Thus the subscripts DON'T match for "Introverts" because of the categorical naming error. Unfortunately, certified counselors are indoctrinated in such categorical confusion, which goes away once decoupled variables replace the coupled ones.

In Example 6, $C = I$ and $B = P$ select $D =$ *I*j in Table 7.1, whereas in Example 5, *E* and *J* choose $D =$ *E*j. These determinations, which precede application of any of the *TD* rules, are entirely categorical and entirely correct. No comparisons or computations are needed.

7.3.3 Subsidiary Decoupled Attitude

Next let *d* represent the *subsidiary* decoupled attitude in the other domain, indicated quantitatively by the smaller decoupled attitude score. Its determination will be comparative rather than categorical, for it examines only the order of the coupled attitude scores *B* and *C*, not their detailed numerical scores. Since *d* must

Table 7.1 Coupled and decoupled attitudes

	$B = P$	$B = J$
$C = E$	***E*p**	***E*j**
$C = I$	***I*j**	***I*p**

Coupled attitudes are shown italic. Decoupled attitudes are shown bold certainly be in the domain opposite the principal one, only its attitude needs to be determined. Consequently modify the notation slightly, using *d* if its attitude is the same as that of the principal *D* but employing *d*′ when the two attitudes differ.

At this point experienced type dynamicists might contend that the *TD* "Attitude Balance" rule allows only the opposite attitude *d*′. As will be discussed more fully in the next subsection, this rule is not in fact universal, being based on an unintentional misreading of Jung. So for the sake of argument, temporarily allow the possibility of either case and learn how to tell which one applies in a particular situation.

Table 4.1 in Sect. 4.5 gives formulas derived there for the decoupled attitudes by Platonic analysis. The detailed quantitative expressions are not relevant at the moment, but they easily yield the comparisons given in Table 7.2 relating the decoupled attitudes to the coupled ones. The primes (′) denote opposite variables, e. g., $E' = I\ (= -E)$ and so on. To put the principal decoupled attitude into the upper left cell, the attitudes in the table are the abstracted versions *B, C, d, D* rather than the specific ones of Table 7.1. Then the two subsidiary possibilities appear in the upper right (*d*) and lower left (*d*′) cells, the remaining cell (*D*′) being irrelevant although shown.

Only the positive entries are of interest. There are two cases. If $B \geq C$, then the subsidiary attitude *d*′ is **opposite** to that of the principal attitude *D*. Thus in Example 6, in which $B = 60\%P > 20\%I = C$, $D = I$j implies $d' = E$p (>0), the subsidiary attitude differing from the principal one as indicated coincidentally by the *TD* attitude balance rule. This subsidiary attitude is in the same COLUMN as the principal attitude. Loomis (1991) calls this situation "ambiversion", attributing the label to various researchers.

On the other hand, if $C > B$, the subsidiary attitude *d* is the **same** as that of the principal attitude. Thus in Example 5, in which $C = 100\%E > 60\%J\ (= B)$, $D = E$j implies $d = E$p, the subsidiary attitude being the same as that of the principal one, contradicting the *TD* attitude balance rule. This time the subsidiary attitude is in the same ROW as the principal attitude. In Sect. 4.7 this situation was defined as "doubleversion". In the author's experience with student design teams, doubleversion occurred between 10 and 20% of the time. Its detection has been valuable for assigning students to appropriate roles on their teams. Many executives are double extraverts, and many good researchers are double introverts.

So far the procedure has been entirely categorical except for the simple numerical comparison of the coupled attitudes. For future reference this will be called the "Attitude Comparison" step. It must be understood that the attitudes being compared are the original decoupled ones *C* and *B* of Jung and Myers&Briggs rather than the new decoupled attitudes.

Table 7.2 Generic coupled and decoupled attitudes

	B	*B'*
C	$\boldsymbol{D = (B + C)/2}$	$\boldsymbol{d = (-B + C)/2}$
C'	$\boldsymbol{d' = (B - C)/2}$	$\boldsymbol{D' = (-B - C)/2}$

Coupled attitudes are shown italic. Decoupled attitudes are shown bold

An important improvement is that the new decoupled attitudes do not suffer from any categorical errors in logic. Table 4.1 shows how to assign numerical scores to the decoupled attitudes, but doing this will wait until it is time to determine any subsidiary modes (not attitudes) overlooked by Type Dynamics. By the way, the suppression of the scores here is an example of categorical information loss. Thus *TD* can sometimes be rehabilitated by recovering some of this information as needed.

7.3.4 TD Rule 1: Attitude Balance

Presently the attitude balance rule is firmly entrenched as a hard and fast Type Dynamics assumption because of an easily misunderstood statement of Jung. Taken from *Psychological Types*, Chap. 10, Sect. *d. The Principal and Auxiliary Functions,* this notorious quotation from lines 668–670 follows:

> Experience shows that the secondary function is always one whose nature is different from, though not antagonistic to, the primary function.

Although this occurs in a three-page section entirely devoted to functions, not attitudes, in which the word "attitude" is never mentioned, it has been quoted by Myers and several others as justifying **attitude** balance. This argument is not valid, however, because Jung's statement was taken out of his context of functions and mistakenly applied to attitudes as well. Notice that even in the statement itself the word "attitude" is not used, although "function" appears twice.

Thus the contradiction between the attitude balance rule and the quantitative analysis of *MBTI* data (e. g., Example 6 in Sect. 7.3.3) arises entirely from taking Jung's correct statement out of its intended context. Realization of this mistake is what first lead the author to stop using Type Dynamics when forming and organizing teams. For better or for worse, that was the start of the current effort to clean up Jung's personality theory by quantifying it.

One can even speculate that if Reynierse and Harker (2008) had unscientifically excluded from the sample the data that incorrectly employed attitude balance, they would have experienced a better match between measurement and prediction, whether or not one statistically acceptable. This should be a source of some comfort to *TD* advocates, for it suggests that, to use an automotive metaphor, one doesn't need to sell the car to get a flat tire fixed. *MBTI*ers then can warmly thank Reynierse for getting the car into the repair shop.

7.3.5 Dominance and Principality

The procedure so far has been largely but not entirely categorical, being augmented by a single simple comparison of the original coupled attitude scores.

In the same vein, another categorical step can determine the dominant mode, subject to another simple comparison, this time of the original function scores.

The two examples show what happens and what to do. In the well-behaved Example 6, the principal decoupled attitude is now known to be $D = I$j, introverted judgment. The corresponding j-function score is 60%T, which definitely exceeds the p-function score 20%N. Thus it is reasonable to regard the combination of the judgment attitude Ij and function T into the cognitive mode Ti introverted thinking as the *dominant* mode. Similar reasoning identifies Ne as the *auxiliary* mode, since it combines the subsidiary decoupled attitude Ep with the p-function N, which has the lower function score.

Things are not so simple in Example 5, deliberately chosen to be a counterexample to conventional Type Dynamics. This time the principal decoupled attitude is Ej extraverted judgment, the subsidiary one being the doubleverted Ep extraverted perception mode. The respective function scores are 20%F and 60%S, whose order is inconveniently opposite to that of the decoupled attitudes. As discussed in Sect. 6.6, this generates an ambiguous dominance relation, making it unwise to characterize either the principal j-mode Fe or the principal p-mode Se as "dominant" The functions in general can be used in this way as a comparison test to tell whether the dominance is clear or ambiguous. For future reference, this test will be known as "Function Comparison". In the troublesome Example 5, one may speak of a dominant j-attitude Ej and a dominant p-function S, but not of a dominant cognitive mode.

This concludes the discussion of applying categorical reasoning to the analysis of *MBTI* data. Aided by two simple comparisons of raw data scores, the first between attitudes and the second between functions, such reasoning can correctly identify both dominant and auxiliary modes in unambiguous situations. Moreover, the two comparisons can steer the analyst away from ambiguous cases, something unaided *TD* can't do. The next section looks more directly at the *TD* rules, where they go astray, and what might be done when that happens.

7.4 Other Type Dynamics Rules

Type Dynamics has provided other rules in addition to "attitude balance" for analyzing the *MBTI* categorically that carry with them implicit simplifying assumptions needed to fill in the information lost by categorization. These additional categorical rules follow for critical examination. Often presented as fundamental principles, they really are only approximations based on simplifications that unfortunately have realistic exceptions.

In terms of the comparison tests of the preceding section, the two unstated *TD* simplifications are first that the person under study is an ambivert ($B \geq C$) and second that its dominance relations are definite. Doubleverted and ambiguous dominance cases are thus automatically excluded from the current *TD* theory, which is uncorrected by the comparison tests.

The use of the *TD* rules following is described on p. 30 of the *MBTI Manual* (3rd edition). To expose their limitations, an example (6) and a counter-example (5) is presented for each of them.

7.4.1 TD Rule 2: The Myers–Briggs Attitude Determines the Domain of the Extraverted Mode

TD says that the *Myers–Briggs* coupled attitude *B*, either *P* or *J*, indicates the domain having the only extraverted mode. Thus in well-behaved *INTP* Example 6, this rule asserts that the *MB* attitude *P* points to the *P*erception domain as containing the extraverted mode, which in this example must be *N*e. That it is the *only* extraverted mode follows from the unstated assumption that the subject is ambiverted, a condition that also identifies the only *j*-mode *T*i.

In *ESFJ* Example 5 this rule would say that the *j*-domain contains an extraverted mode *F*e, which indeed it does. But the subject here is a double extravert, so *F*e cannot be the only extraverted mode. Clearly the *p*-function must also be extraverted, producing the mode *S*e. This will also cause trouble with Rule 3 to follow.

7.4.2 TD Rule 3: The Dominant Mode and Jung Attitudes are the Same

Rule 3 assumes that the Jungian coupled attitude *C*, either *E* or *I*, is also the attitude of the dominant mode. This works fine in *INTP* Example 6, since Rule 2 combines with the subject's ambiversion to confirm *T*i as the sole introverted mode. To verify that *T*i is dominant, recall that its attitude *I*j was established as the principal decoupled attitude in Sect. 7.3.2. Moreover, its *j*-function score 60%*T* certainly exceeds the *p*-function score 20%*N*, a circumstance consistent with the successful function comparison assumed in advance. Together these two facts prove that the combination of *I*j with *T* forms the dominant mode *T*i.

Things don't work out for the contrarian *ESFJ* Example 5, however. Although Rule 3 does say that the dominant mode (if there is one) will be extraverted, this time there are *two* extraverted modes due to doubleversion. And trying to determine which one is dominant leads to an impasse. This is because although *E*j was found in Sect. 7.3.2 to be the principal decoupled attitude, the corresponding *j*-function score 20%*F* is less than the *p*-function score 60%*S*. Thus this example is troubled by ambiguous dominance, making Rule 3 inapplicable here.

Notice that all three of the numbered *TD* rules can easily be replaced by the categorical determinations in Sects. 7.3.2 and 7.3.3 of the principal and subsidiary decoupled attitudes, as long as they are accompanied by attitude and function comparisons. Fortuitously the *TD* rules do get things right much of the time, as in Example 6. This has led traditional practitioners to defend *TD* even in the face of the unfavorable statistical studies of Reynierse and Harker (2008) and clear counterexamples like Example 5. But why tolerate even a small amount of unreliability if it can be avoided easily?

7.4.3 TD Rule 4 (Unstated): All Personalities are Bimodal

The first three *TD* rules led to errors that the preceding section has shown to be easily avoidable. The fourth rule to be discussed here may not be recognized by experienced *MBTI* users because it has never before been stated explicitly. It is that all personalities are *bimodal*, having no cognitive modes except the dominant and auxiliary. Although this notion has never been expressed in so many words, it is implicit in the *TD* procedure because it stops after applying the first three rules.

The earlier part of this book showed that, contrary to this unstated "rule", there can be as many as two more significant modes, one "subsidiary" mode in each of the two domains. As will be shown by Example 6, these subsidiary modes may sometimes have insignificantly low scores and so be discarded as negligible. The other possibility though, illustrated by that pesky Example 5, is quite exciting. The subsidiary modes may be significant enough to bring to light important parts of a personality which would be overlooked by conventional *TD* analysis, even as corrected in the preceding section. Such personalities will be called *multimodal* when they have more than two modes.

This result is of great potential importance for the identification, education and unleashing of human potential. Revelation of these subsidiary modes, which have already been harnessed for constructing improved design teams, should in particular interest school counselors because they provide career guidance to high school and college students.

All this has been made quantitative in the earlier part of the book, particularly in Chap. 6. This however would require some calculation, which although simple, will be circumvented here, somewhat in the spirit of existing *MBTI* certification programs. However, one calculation is needed now to reflect the newer paradigm of decoupled attitudes. After all, if the data comparisons introduced to shake out the errors in the old *TD* rules were useful, why not bring in scores for the decoupled attitudes as well, especially since their formulas are so elementary?

The arithmetic goes back to Table 4.1, the quantitative forerunner of Table 7.1 used to determine the principal and subsidiary decoupled modes. One merely needs to evaluate them to obtain the associated scores. Thus in Example 6, the decoupled attitudes are $D = [(I + P)/2]I\text{j} = [(20\%I + 60\%P)/2]I\text{j} = 40\%I\text{j}$ for the principal attitude, and $d = [(E + P)/2]E\text{p} = [(-20\%I + 60\%P)/2]E\text{p} = 20\%E\text{p}$ for the subsidiary attitude.

In Example 6 the principal cognitive modes have already been determined to be the dominant *T*i and the auxiliary *N*e. In each domain a simple comparison of the attitude to the function immediately shows if there is also a subsidiary mode to consider. If, as in the example, the scores are equal for the *p*-attitude 20%*E*p and *p*-function 20%*N*, there will be no subsidiary attitude. Even when the difference is small, as for 40%*I*j and 20%*T*, the subsidiary mode may be insignificant (here 10%*F*i) and not worth considering. Thus Example 6 happens to fulfill *TD*'s rule 4 neglecting subsidiary modes.

But, as in the ever-contrary Example 5, a notable difference between attitude and function signals a subsidiary mode. When the function exceeds the attitude, the subsidiary mode will have the same function as that of the principal mode. In Example 5 this happens in the *p*-domain where 60%*N* > 20%*E*p, indicating the presence of intuitive subsidiary mode *N*i (score 20%). Similarly, when the attitude exceeds the function, the subsidiary mode will have the same attitude as that of the principal mode. Thus in Example 5's *j*-domain, 80%*E*j > 20%*F*, clearly indicating an extraverted subsidiary mode *T*e scored at 30%.

A Platonic ideal *MBTI* type always has a subsidiary mode in its auxiliary domain. For Example 6's *INTP* type the Platonic ideal (100%*I, N, T, P*) would have a principal decoupled attitude of 100%*I*j generating a dominant mode *T*i with no subsidiary *j*-mode. Its 0%*E*p subsidiary decoupled attitude would indicate an auxiliary mode *N*e having the same 50% score as its subsidiary mode *N*i. Section 6.5 discusses these trimodal Platonic ideal types further.

Although non-numerical comparison is often good enough to detect or rule out subsidiary modes, the small extra effort needed to compute the scores is usually justified. It is hoped that this subsection will inspire type counselors everywhere to re-examine their files to find the human potential they may have overlooked.

7.5 The Type Table

Table 7.3 exhibits the $2^4 = 16$ combinations of the four *MBTI* letter pairs. Believing at the time in the validity of the first three *TD* rules, in particular attitude balance, Myers assumed universal ambiversion, which if true would have given

Table 7.3 Ambivert type table with cognitive mode assumptions

	ST	*SF*	*NF*	*NT*
I_J	*ISTJ*	*ISFJ*	*INFJ*	*INTJ*
	*S*i, *T*e	*S*i, *F*e	*N*i, *F*e	*N*i, *T*e
I_P	*ISTP*	*ISFP*	*INFP*	*INTP*
	*T*i, *S*e	*F*i, *S*e	*F*i, *N*e	*T*i, *N*e
E_P	*ESTP*	*ESFP*	*ENFP*	*ENTP*
	*S*e, *T*i	*S*e, *F*i	*N*e, *F*i	*N*e, *T*i
E_J	*ESTJ*	*ESFJ*	*ENFJ*	*ENTJ*
	*T*e, *S*i	*F*e, *S*i	*F*e, *N*i	*T*e, *N*i

the 16 dominant and auxiliary mode pairs shown in Table 7.3. For each type, the dominant mode is shown first, the auxiliary mode following it.

Being unaware at the time of the other 16 doubleverted combinations shown in Table 7.4, Myers called the 16 four-letter combinations "personality types". In this book this limited designation will be preceded either by the letters "*MBTI*" or "ambiverted" to distinguish it from that for the "doubleverted" types of Table 7.4. Table construction details are omitted here, but type descriptions will be discussed in the subsection following.

7.5.1 Myers Type Descriptions

Myers based her bimodal type descriptions only on the ambiverted dominant and auxiliary modes assumed by Type Dynamics and detailed in Table 7.3. A professional writer, she compactly combined key words and phrases associated with the dominant and auxiliary modes, using about twice as many dominant words as auxiliary ones. This is illustrated here by her ambiverted *ENTP* description (*Manual 2*, p. 21) combining dominant ***Ne*** (**boldface**) and auxiliary *Ti* mode descriptions.

> **Quick, ingenious, good at many things. Stimulating company, alert and outspoken.** May argue for fun on either side of a question. **Resourceful in solving new and challenging problems, but may neglect routine assignments. Apt to turn to one new interest after another.** Skillful in finding logical reasons for what they want.

Ironically, unintentionally leaving out fully half of the 32 four-letter personality types may at the time have been beneficial to the then novel practice of typology. If 16 types were daunting enough, 32 might have been fatal for acceptance of the new theory. That the budding craft survived such a large oversight is probably because there aren't that many doubleverts around. Among the Stanford sophomores taking the Teamology seminar, less than 20% could be considered doubleverts. But recognizing doubleversion at long last is certainly the right thing to do, for the sake not only of scientific accuracy, but also for reducing the number of dissatisfied clients, some of whom write nasty attack articles in the popular press.

Table 7.4 Doublevert type table with cognitive mode assumptions

	ST	*SF*	*NF*	*NT*
*I*_J*	*I*STJ*	*I*SFJ*	*I*FNJ*	*I*NTJ*
	Si, *Ti*	*Si*, *Fi*	*Ni*, *Fi*	*Ni*, *Ti*
*I*_P*	*I*STP*	*I*SFP*	*I*NFP*	*I*NTP*
	Ti, *Si*	*Fi*, *Si*	*Fi*, *Ni*	*Ti*, *Ni*
*E*_P*	*E*STP*	*E*SFP*	*E*NFP*	*E*NTP*
	Se, *Te*	*Se*, *Fe*	*Ne*, *Fe*	*Ne*, *Te*
*E*_J*	*E*STJ*	*E*SFJ*	*E*NFJ*	*E*NTJ*
	Te, *Se*	*Fe*, *Se*	*Fe*, *Ne*	*Te*, *Ne*

7.5.2 A Table for Doubleverts

Table 7.4 shows how a type table for doubleverts might be organized in terms of dominant and auxiliary modes having the same attitude. It was constructed from Table 7.3 simply by reversing the attitudes of the auxiliary modes. An asterisk (*) follows the first type code letter as a reminder that for doubleverts the Jung *E–I* score always exceeds the *MB P–J* score. Construction of the appropriate type descriptions is left as a potentially profitable exercise for the cottage industry of writers about type. Remember though that because of additional subsidiary modes, even the new table would not complete the task of describing a personality. Until doublevert descriptions are published, one must be content with reading the descriptions of all the significant modes, as many perhaps as the four in Example 5.

7.5.3 Function Scores

Recall that in addition to the coupled attitude comparison for ambiversion, one needs to see if the function with the larger score is in the same domain as the principal decoupled attitude. Whenever this comparison fails, the function with the larger score will have its letter marked with an asterisk in the type code to warn of the anomaly.

Consider Example 7: 20%*E*, 30%*N*, 60%*T*, 60%*P* in which the function scores of ambivert *ENTP* Example 1 have been interchanged. The letter code for Example 7 will thus be *ENT*P*, indicating to type dynamicists that dominance is ambiguous here, *T*'s score exceeding that of *N*'s even though the principal decoupled attitude is in the *p*-domain. Mode Table 7.5 shows this ambiguity, the principal *j*-mode score exceeding slightly that of the principal *p*-mode. This example has a significant *extra*verted subsidiary *j*-mode 20%Te, whereas Example 1 had a significant *intro*verted subsidiary *p*-mode 30%Ni. Ambiverted Type Table 7.3 could be slightly misleading for an *ENT*P*, whose principal modes are closer to those of the

Table 7.5 Mode table for *ENT*P* Example 7

Perception			Judgment		
	S	*N*	–	*T*	*F*
		30% iNtuiton (Concepts)		60%* Thinking (Things)	
*E*p 40%# Exploration	***Se*** **5% Experiment**	***Ne*** **35%# Ideation**	– *E*j	***Te*** **20% Organization**	***Fe***
*I*p	***Si***	***Ni***	– *I*j 20% Appraisal	***Ti*** **40%* Analysis**	***Fi***

ambiverted *INTP* entry. Such tabular complexity can be avoided by using the mode descriptions directly instead of the type descriptions.

It is reasonable to use two asterisks when necessary as in doubleverted Example 5. There the "starified" type code would be *E*S*FJ*, alerting the typologist to the perversity of mode Table 6.6

Since each asterisk increases the number of type codes by 16, there appear to be at least 64 four-letter types, even without accounting for subsidiary modes. Although the starred types do not occur often, such personalities deserve to have their uniqueness recognized, especially in the context of team membership or career choice. The asterisks also remind the personologist when care needs to be exercised with the overly restrictive rules of traditional type dynamics.

7.5.4 The End of Quantitative Type Analysis

This is as far as quantitative reasoning can go with Jung's personality theory. It has not only generated a consistent way to identify the main cognitive modes associated with any set of *MBTI* scores, but also shown how to repair the incomplete categorical approach of Type Dynamics.

There remain, however, further categorical conclusions using the cognitive modes already determined to infer things about those modes remaining. To prevent the transmission of *TD* errors to these other modes, more study is needed. It will begin with the next section's examination of the two modes opposite the dominant and auxiliary modes. This will set the stage for studying, in the chapter following, all the other modes as well in the context of Beebe's Archetypal Type Theory. None of this will involve quantitative analysis, only careful categorical reasoning.

7.6 Dominant and Auxiliary Opposite Modes

Now that as many as four modes have been scored positively, what about those remaining? Life does not confine itself to preferred situations. How for instance would a *T*e Organizer be expected to react to an introverted feeling *F*i situation requiring moral evaluation? This section will examine two modes in addition to those already highlighted, leaving the other four for the next chapter. The reasoning will be entirely categorical, the problem being to start with the right categories in the first place.

This section will only consider the modes opposite the dominant and auxiliary ones. Whenever the dominance is ambiguous, the reasoning to follow must be used cautiously and confirmed by the analyst's clinical experience.

Jung himself saw the mode opposite the dominant one as important enough to merit a name: *anima* for males and *animus* for females. In Latin, meaning "spirit", *anima* is feminine and *animus* masculine, genders opposite to that of the person. For Jungians this archetypal figure always stands for the complementary, contra-sexual side of the psyche, representing, in the words of Jacobi (p. 114), "the image of the other sex that we carry in us as individuals and also as members of the species". Understanding this gateway to the unconscious shadow is important to analyst and patient alike in Jungian psychotherapy for personal individuation, the process of transcending the ego and the shadow. Since as noted in Chap. 3 the dominant mode is relatively easy to determine even without a questionnaire, its opposite anima/us mode is equally evident when the dominant mode is unambiguous.

In 1983 Grant, Thompson, and Clarke sought to characterize the mode opposite the auxiliary, an obvious extension of Jung's anima/us concept. This designation, widely promulgated by Brownsword (1987), is of course entirely categorical, its correctness depending only on getting the auxiliary mode right. As was typical in those days, all people were assumed to be ambiverts with the auxiliary attitude always opposite to the dominant attitude. This of course would be incorrect for a doublevert. So Reynierse and Harker's (2008) statistical study of Type Dynamics naturally found no correlation of the properties predicted by Grant and Brownsword for any mode opposite to a mode incorrectly assumed to be auxiliary.

7.6.1 In Defense of Grant, Brownsword and Beebe

One suspects that separating the data into ambiverted and doubleverted groups would have shown acceptable correlations. This scientifically questionable statement is made in defense of Grant and Brownsword, whose simple categorical analysis is hard to dispute. All they need to make their predictions work are the correct auxiliary modes.

To foreshadow Chap. 8: Beebe gave archetypal names and descriptions to all eight modes, of which the most conscious "Ego" four considered in this section are given in Table 7.6. The "levels" are related to the order of the mode scores, "1" referring to the mode with the highest positive score and "2" to the corresponding auxiliary mode. Lower levels are assigned categorically, 3 going to the auxiliary's opposite and 4 to the dominant's opposite. Chap. 8 will give a more complete discussion of these, along with the other four "shadow" modes.

Table 7.6 Ego archetypes

Level	Archetype	Mode Identification	Description
1	Hero/Heroine	Dominant	Masterful
2	Parent: Father/Mother	Auxiliary	Helpful
3	Boy/Girl: Puer/Puella	Aux. Opposite	Playful
4	Spirit: Anima/us	Dom. Opposite	Spiritual

Not allowing for doubleversion, Beebe's predictions also failed the statistical study of Reynierse and Harker. However, with the current chapter's corrections of Type Dynamics, the categorical assignment of the RIGHT modes to the archetypes should work fine once the dominant and auxiliary modes are correct.

7.6.2 An Ego Mode Example

This theory will be illustrated using ambiverted *ENTP* Example 1, in which *N*e is dominant and *T*i auxiliary. There is also a subsidiary mode $\underline{N\mathrm{i}}$ involved. Suppose the subject is male. The dominant mode archetype *N*e (Ideator) is the Hero. Combining mode and archetype keywords gives the dominant mode archetype as the "Heroic Ideator". Similarly the auxiliary mode *T*i archetype is Fatherly Analyst. The opposite dominant mode *S*i Anima archetype is "Spiritual Scholar", while the opposite auxiliary mode *F*e archetype is Boyish Communitarian. These simple four 2-word descriptors are not a bad introduction, either to self-knowledge on the road to individuation, or to self description for making conversation with a new partner or team-mate. The remaining subsidiary "positive shadow" mode $\underline{N\mathrm{i}}$ will be considered, along with the three remaining shadow modes *T*e, *F*i and *S*e, in the next chapter.

7.7 Preference Multidimensionality

Section 7.2 honored the pioneers whose enthusiasm and enterprise launched the *MBTI* and spread it around the world in the last half of the 20th century. Experimental psychologists Reynierse and Harker (*R&H*) also deserve credit for their careful 2008 statistical and logical study showing that many *TD* attitude predictions did not fit the data. Their work confirms the order Jung conceived for the functions. That is, the most powerful dominant function was indeed supported in the opposite domain by a secondary auxiliary function whose opposite is of tertiary importance, leaving the fourth or "inferior" position to the mode opposite the dominant. It was the assignment of the attitudes to the functions by *TD* that was found not to fit the data.

R&H did find, however, that groups of *MBTI* variables predicted personality characteristics quite well. They call this approach to typology "preference multidimensionality" (*PM*). The relations they found were always strictly proportional, what mathematicians would call "linear". This is particularly interesting from the standpoint of this book, whose main quantitative relationships are *linear* interpolations—never extrapolations—of Platonic idealizations.

R&H worked with the *MBTI* variables in various combinations rather than combining them into cognitive modes. Moreover, they consider Jung's partition of modes into perception and judgment domains unnecessary, although not incorrect.

This book has stayed with Jung's two-domain concept in order to discuss his personality theory. *Teamology* independently found this partition useful for describing team roles. It is also valuable for general function (-attitude) analysis. Reynierse also exposed categorical errors in *TD*'s logic that have been fixed in this book by the substitution of decoupled attitudes for the old coupled ones. Decoupling the attitudes may also improve the *PM* correlations.

Reynierse and Harker performed a great service to *MBTI* users simply by showing that Type Dynamics was off the rails and needed repair. It motivated the current chapter's intention to bring *MBTI* theory and practice into the 21st century.

7.8 Concluding Summary

This chapter began by acknowledging the pioneering efforts of Isabel Myers in getting the *MBTI* into wide circulation. It then reviewed Reynierse's exposure of the categorical naming error of the original coupled attitudes, recalling that attitude decoupling automatically clarifies and corrects the situation. Four categorical *TD* assumptions were described, and two simple data comparison tests, one for doubleversion and one for dominance ambiguity, were presented to detect when, and when not, the *TD* assumptions hold. Counterexamples to all four assumptions were given that suggest why the Reynierse and Harker study did not find a match between *TD* theory and the data.

The impact of the *TD* errors on the *MBTI* Type Table were then described. The two comparison tests were shown to generate 48 more "types", even without considering subsidiary modes. This ended the quantitative study of Jung's personality theory, leaving some more categorical matters to be examined.

Then the archetypal theories of Grant, Brownsword and Beebe concerning the modes opposite to the dominant and the auxiliary modes were discussed. Discredited by the Reynierse and Harker study, they were defended as reasonable categorical concepts unintentionally based on correctable *TD* errors that had occasionally generated invalid dominant and auxiliary modes. It was suggested that such an archetypal approach, being entirely categorical, should be validated clinically rather than quantitatively. The non-quantitative archetype theory for the other four "shadow" modes will for completeness be outlined in the next chapter.

The chapter ended with a salute to the meticulous and difficult work of Reynierse and Harker, whose courageous efforts alerted the profession to the repairs needed to *MBTI* theory. The fixes should bring out the instrument's previously unseen power and propel it into an expanded and productive future.

7.9 Exercises

7-1 Using your results on the teamology questionnaire:

a. Find your *TD* dominant and auxiliary modes using only the type code.
b. Which, if any, of the two comparison tests would you pass?

c. Quantitatively determine all your significant modes.
d. As in Sect. 7.6.2, combine mode and archetype keywords to make descriptors for your first four ego modes.

7-2 If you have *MBTI* scores:

a. Find your *TD* dominant and auxiliary modes using only the type code.
b. Which, if any, of the two comparison tests would you pass?
c. Quantitatively determine all your significant modes.
d. As in Sect. 7.6.2, combine mode and archetype keywords to make descriptors for your first four ego modes.

7-3 Team-mate *X* of the student in Example 5 has scores 20%*I*, 100%*N*, 20%*F*, 20%*P*.

a. Find the *TD* dominant and auxiliary modes using only the type code *INFP*.
b. Which, if any, of the two comparison tests does this student pass?
c. Quantitatively determine all significant modes for this student.
d. As in Sect. 7.6.2, combine mode and archetype keywords to make descriptors for the first four ego modes.

7-4 Team-mate *Y* of the student in Example 5 has scores 60%*I*, 100%*S*, 100%*T*, 80%*J*.

a. Find the *TD* dominant and auxiliary modes using only the type code *ISTJ*.
b. Which, if any, of the two comparison tests does this student pass?
c. Quantitatively determine all significant modes for this student.
d. As in Sect. 7.6.2, combine mode and archetype keywords to make descriptors for the first four ego modes.

7-5 Team-mate *Z* of the student in Example 5 has scores 60%*E*, 20%*S*, 20%*F*, 60%*P*.

a. Find the *TD* dominant and auxiliary modes using only the type code *ESFP*.
b. Which, if any, of the two comparison tests does this student pass?
c. Quantitatively determine all significant modes for this student.
d. As in Sect. 7.6.2, combine mode and archetype keywords to make descriptors for the first four ego modes.

7-6 On a judgment mode graph similar to Fig. 5.2 in Sect. 5.4.3, plot the *j*-modes for Example 5 and his three team-mates *X, Y*, and *Z*.

Chapter 8
Shadow Archetypes

Who knows what evil lurks in the hearts of men?
–"The Shadow" radio character,
played by Orson Welles c. 1940.

8.1 Introduction

Earlier chapters confined themselves to the psyche's two conscious cognitive mode pairs, comprising what Freud and Jung described as the "ego". The present chapter will discuss the remaining four modes associated with the unconscious "shadow" as Jung called it.

Sect. 7.6 discussed the two categorically defined *anima/us* and *puer/puella* ego modes respectively opposite the dominant Hero/ine and auxiliary Father/Mother modes. Their characters were determined by clinical experience rather than quantitative statistical study. The present chapter will discuss Beebe's extension of this clinical approach to the four remaining shadow modes of the unconscious. His theory, being entirely categorical, can only be right if the dominant and auxiliary modes are correct. Beebe's extension will be named the "bimodal archetype theory" because in it all six remaining modes are defined categorically in terms of the two dominant and auxiliary modes. His theory does not include any of the subsidiary modes discussed in Chap. 6.

The chapter begins by describing this bimodal theory, adapted here for informal use by *MBTI* practitioners taking account of the quantitative type dynamics repairs of the preceding chapter. Then the theory is extended to the multimodal situations, with each subsidiary mode treated as a "positive shadow". This concept of Jung's described by his disciple Jolande Jacobi, leads to a novel multimodal approach overlaying the bimodal theory with subsidiary positive shadow interpretations.

Another approach concerns asymmetric cases in which one modal variable has a much larger score than the other. This is the circumstance that generates subsidiary modes. It is convenient then to split the modes into "roles", a *Teamology* strategy for organizing teams. The role formulation can help a team troubleshoot a problem with a particular activity such as prototyping.

These concepts are illustrated with numerical and tabular examples. In addition to tri- and quad-modal examples, there is a two-person and a team quartet situation. Although based on the quantitative reasoning of the earlier parts of the book,

D. J. Wilde, *Jung's Personality Theory Quantified*
DOI: 10.1007/978-0-85729-100-4_8, © Springer-Verlag London Limited 2011

the conclusions are meant only to be suggestive, in practice offering guidance to counselors, managers and therapists with other situational information to factor in.

8.2 A Bimodal Version of Beebe's Archetype Model

This section develops Beebe's archetype model (Beebe 2007; Harris 1996, pp. 65–75), which is supported anecdotally by clinical and personal experience rather than quantitatively by questionnaire data. Suppose the dominant and auxiliary modes are known. Beebe's model is here called "bimodal" because it does not take into account any "subsidiary" modes.

Reynierse (2009) has objected that the errors of Type Dynamics undermine this archetypal approach. But as discussed in Sect. 7.6.1, the model should be all right if it is based on the correct Platonic dominant and auxiliary modes. Henceforth Beebe's archetypal names "Hero/ine" and "Parent (Father/Mother)" respectively will be used for the dominant and auxiliary modes.

Recall that the archetypal model has two other modes aside from the Hero and Parent. They are the "Anima/us" opposite the Hero and the "Puer/Puella (Child)" opposite the Parent, for a total of four ego modes.

8.2.1 The Shadow Archetypes

To define the four modes remaining, which for Beebe are all in the shadow, define the operation of *contraversion*, which reverses the attitude of a mode. The resulting mode, said to have been "contraverted", is called the "contravert" of the original mode. For instance, the contravert of extraverted feeling is introverted feeling.

Table 8.1 gives the archetype identifications for all eight modes, together with their informal descriptions. For the reader's convenience, the top four lines are repeated from Table 7.6 in Sect. 7.6.1. These lines echo the numerical order of the

Table 8.1 Archetype definitions

Level	Archetype	Mode identification	Description
1	Hero	Dominant	Masterful
2	Parent	Auxiliary	Helpful
3	Child (Puer/Puella)	Auxiliary opposite	Playful
4	Anima/Animus	Dominant opposite	Spiritual
5	Opposer	Dominant contravert	Undermining
6	Witch/Senex	Auxiliary contravert	Obstructive
7	Trickster	Witch/Senex opposite	Deceptive
8	Daemon	Opposer opposite	Dangerous

ego mode scores: "1" for the highest positive, down to "4" for the most negative. As Table 8.1 shows, however, that shadow levels 5 through 8 are assigned categorically by contraversion and/or opposition rather than by their questionnaire scores.

Beebe's terminology for the shadow modes includes some words that are perhaps unfamiliar. "Senex", the male counterpart of the feminine Witch, is Latin for "old man", the root of "senile". The dictionary meaning of "daemon" is

> (in ancient Greek belief) a divinity or supernatural being of a nature between gods and humans
>
> an inner or attendant spirit or inspiring force.

New keyword descriptors for these terms, as well as for "Opposer" and "Trickster", are in the Description column of Table 8.1.

It is of course superficial to describe each archetype with only a single keyword. This is done to avoid giving the impression that the chapter intends to advise therapists and counselors how to work with their clients' shadows in detail. Instead, the keywords are merely aids to following and emphasizing the categorical nature of the archetype theory, the true subject of the chapter. For deeper descriptions of the archetypes themselves, consult the Beebe (2007) and Harris (1996) references.

8.2.2 A Bimodal Example

Following is a bimodal Example 8, constructed from Example 1 by lowering the intuition score to 40%*N*, which makes the *N*i score vanish. Keeping all other *MBTI* values the same yields the MBTI scores: 20%*E,* 40%*N*, 20%*T,* 60%*P*, which generate decoupled attitude scores of 40%*E*p and 20%*I*j as in Example 1a, and function scores of 40%*N* and 20%*T.* The modes are then 40%*N*e, 0%*N*i; 20%*T*i, 0%*F*i. The corresponding *MBTI* values are *E*6, *N*12, *T*6, *P*18. Thus take *N*e to be dominant and *T*i auxiliary. Table 8.2 maps the example archetypes on to the cognitive modes, and both genders are listed.

Table 8.2 Archetype map for Example 8

Se	*Ne*	–	*Te*	*Fe*
Opposer			Auxiliary	Auxiliary
Opposite	**Dominant**		Contravert	Opposite
				PUER/
DAEMON	**HERO**		WITCH/	PUELLA
	/INE		SENEX	(CHILD)
Si	*Ni*	–	*Ti*	*Fi*
Dominant	Dominant		**Auxiliary**	Witch/Senex
Opposite	Contravert		**FATHER**	Opposite
ANIMA/			**/MOTHER**	
ANIMUS	OPPOSER		**PARENT**	TRICKSTER

If the subject is male, as in the ego mode example 1 in Sect. 7.6.2, the dominant mode archetype for *ENTP* Example 8 is the (1) Heroic Ideator. Similarly the auxiliary mode archetype is the (2) Fatherly Analyst, again as in Example 1. The opposite modes also match those of Example 1: (3) Boyish Communitarian, and (4) Spiritual Scholar. What's new in this chapter are the shadow descriptions: (5) Imaginative Underminer, (6) Organized Obstructor, (7) Deceptive Evaluator, and (8) Dangerous Experimenter.

Just as the archetype and modal keywords are intentionally superficial to avoid taking over the function of therapist or counselor, such combinations of them as "Imaginative Underminer" may seem too compact for plausibility. But a little reflection can make sense of them. An Imaginative Underminer, for instance, could be one who imagines non-existent obstacles to his partner's or team's suggestions or goals. Whether or not such paired keywords are relevant to the situation at hand must be decided on the spot by the counselor or team supervisor.

8.3 Multimodal Archetypology

Consider now the archetypology of multimodal personalities having subsidiary modes along with the dominant and auxiliary modes of Beebe's theory. In terms of the dominant and auxiliary modes, if determined correctly, the subsidiary modearchetypes will be in the shadow according to the bimodal theory. This would appear to contradict the subsidiary mode's positive score, however.

The present section seeks to resolve this conceptual conflict in two ways. In Sect. 8.3.1 the first approach brings in Jung's little mentioned notion of the *positive shadow* occurring when some useful characteristic has been repressed. The second, in Sect. 8.3.2, treats subsidiary modes as expressions of asymmetry that can be reduced by splitting the cognitive modes into roles, a strategy used in team organization (Wilde 2009, Chap. 4). This redefinition in effect rotates the domain axes, replacing the diagonal cognitive mode axes with the rectangular axes of decoupled attitude versus function.

8.3.1 Positive Shadows

The mode subsidiary to the dominant mode is called the "Subsidiary Hero" (*N*i in Example 1b). A subsidiary mode may be either *co-functional*, having the same function as the principal mode, or *co-attitudinal*, having the same decoupled attitude as the principal mode. Similarly the mode subsidiary to the auxiliary mode is called the "Subsidiary Parent".

The bimodal archetype theory would regard subsidiary modes as being in the shadow. But their positive scores, if significant, would argue instead that they

Table 8.3 Subsidiary positive and negative shadows

Level	Negative		Positive	
	Archetype	Description	Archetype	Description
5	Opposer	Undermining	**Dreamer**	**Speculative**
6	Witch/Senex	Obstructive	**Conserver**	**Resolute**
7	Trickster	Deceptive	**Jester**	**Amusing**
8	Daemon	Dangerous	**Daredevil**	**Rash**

belong with the ego, expanded beyond the two modes of the bimodal formulation. Jung seems to have anticipated ambiguous modes of this sort, naming them *positive shadows*. His disciple Jacobi (p. 112), writes the following about this concept:

> But paradoxical as it may seem at first sight, the shadow or "alter ego" may also be represented by a positive figure, for example, when the individual whose "other side" it represents is living "below his level", failing to fulfill his potentialities, for then it is his positive qualities that lead a dark shadow existence.

It is proposed here that in such circumstances the shadow archetypes be replaced by the subsidiary positive archetypes listed in Table 8.3. This idea is not presented as authoritative. It is merely a supposition open to the criticisms and modifications of experienced professional therapists and counselors, who should consider it a initial attempt to fill a gap in the existing archetype theory generated by the newly discovered phenomenon of subsidiary modes.

In Example 1 then, the subsidiary mode introverted intuition *N*i would be described archetypally by the positive "speculative imagination" rather than the bimodal negative "undermining imagination". Usage of the other terms of Table 8.3 will be demonstrated in the examples of Sect. 8.4. Contrast this description with the "masterful imagination" characterization that would have been employed for the dominant first, rather than fifth, archetype level.

This entirely speculative positive shadow idea would seem potentially useful for individual counseling and therapy. The next subsection describes another approach devised principally for interpersonal and team organization problems.

8.3.2 Asymmetry and Roles

Jung's original cognitive mode formulation informally assumes strict symmetry, indeed equality, between the function and the decoupled attitude scores. Equal scores do not produce a subsidiary mode.

There is a construct from Chap. 4 of *Teamology* (Wilde 2009) that has some promise of dealing with, or at least mitigating, the effects of asymmetry. It is the concept of *roles*, essentially a partition of each mode square into two roles expressed as equal right triangular sectors as shown in Fig. 8.1 for Perception and Fig. 8.2 for Judgment. Each role triangle is labeled with the two variables adjacent to it, first with the rectangular coordinate and then with the diagonal coordinate.

For example, in Fig. 8.1 the Ideation mode *N*e is decomposed into two roles, the Entrepreneur *E*p, *N*e and the Innovator *N, N*e. Notice that for a mode square the mode symbol (*N*e in the example) is repeated in its two roles. In similar fashion a large triangular region can be formed from the Innovator *N, N*e and Visionary *N, N*i roles to represent the intuition (Concept) function *N*. Triangular regions for the decoupled attitudes can also be constructed by combining adjacent roles having that attitude. Thus Tester *E*p, *S*e and Entrepreneur *E*p, *N*e combine to make the extraverted perception Explorer attitude *E*p.

The role concept offers a way to understand the archetypal effects of subsidiary modes. A subsidiary mode co-functional with its principal mode can be considered to rotate the principal archetype one role sector toward the subsidiary mode. From this point of view the subsidiary introverted intuition mode *N*i is seen to rotate the Hero archetype into the Visionary *N, N*i role sharing the *N* function with the Innovator *N, N*e role. This generates the function triangle Innovator-Visionary *N* (Concept) as the location of the principal archetype, dominant in this case. The other archetypes are assumed to rotate in the same direction. Thus the Opposer archetype now combines the two *I*p roles Strategist *I*p, *N* and Inspector *I*p, *S* into the coupled attitude triangle *I*p (Focus).

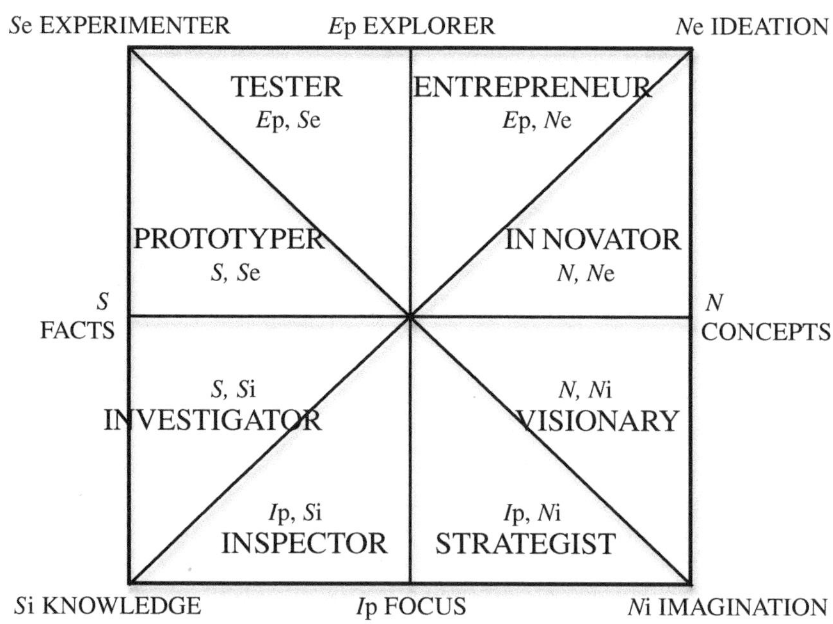

Fig. 8.1 Perception roles

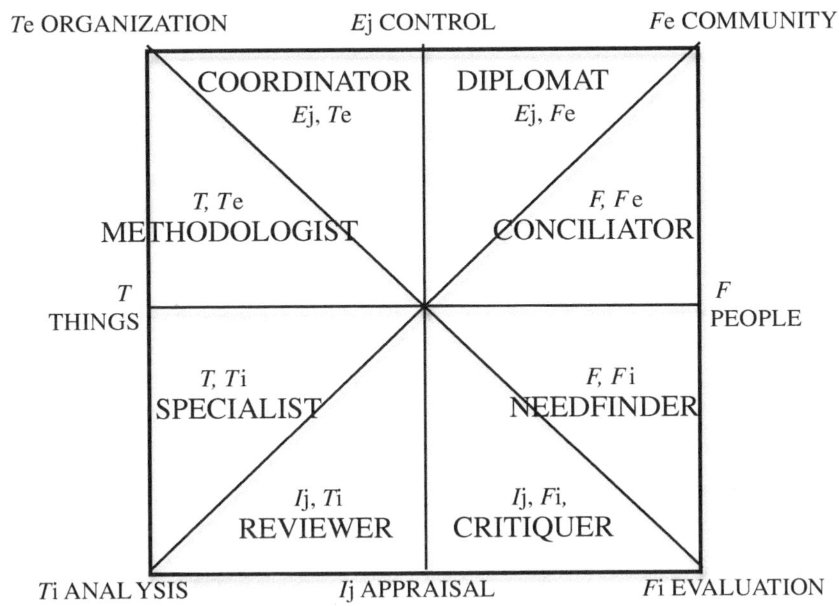

Fig. 8.2 Judgment roles

Table 8.4 Archetype map for Example 1 with a positive shadow

Se	*Ne*	–	*Te*	*Fe*
Opposer Opposite DAEMON	 **Dominant** **HERO**		Auxiliary Contravert WITCH/ SENEX	Auxiliary Opposite PUER/ PUELLA (CHILD)
Si	*Ni*	–	*Ti*	*Fi*
Dominant Opposite ANIMA/ ANIMUS	Dominant Contravert SUBS. HERO ***DREAMER***		 **Auxiliary** **PARENT**	Witch/Senex Opposite TRICKSTER

This approach also works for co-attitudinal subsidiary modes. Example 5 has a subsidiary mode *T*e whose Control attitude *E*j is the same as that of the principal mode *F*e. In this case the Hero archetype would rotate into the Control *E*j attitude triangle covering the two roles Coordinator *E*j, *T*e and Diplomat *E*j, *F*e. The remaining *F*e role Conciliator *F*, *F*e would then share the Opposer archetype with the Needfinder *F*, *I*i role, while the Methodologist role *T*, *T*e remaining from *T*e would join the Specialist *T*, *T*i role to form the eighth level Daemon archetype.

Combining roles to form archetypes is compatible with the *Teamology* method for assigning team members to team duties. Sects. 8.4.2 and 8.5.2 will describe a rather rare situation in which team organization seems to involve a shadow role.

8.4 Multimodal Examples

8.4.1 Trimodal Example 1

Table 8.4 shows how the positive shadow idea applies to trimodal Example 1. The only difference from Table 8.2 is in subsidiary mode *N*i, the Subsidiary Hero mode. Its archetype "Dreamer" is shown in *bolditalic*. In the team role formulation of Sect. 8.3.2, "Visionary" would be Heroic instead of Opposing, and "Entrepreneur" would be "Daemonic" instead of Heroic, as shown in Fig. 8.3

8.4.2 Quadmodal Example 5

Recall the Example 5 scores: 40%*S*e, 20%*S*i; 50%*F*e, 30%*T*e. This is an ambiguous dominance case with two possibilities: the dominant mode is either *F*e or *S*e. If *F*e is dominant, then *T*e is dominant subsidiary and co-attitudinal. Also *S*e is auxiliary, and the auxiliary subsidiary *S*i is co-functional. Table 8.5 displays the male gender archetypes for Example 5 with *F*e dominant.

On the other hand If *S*e is dominant, then *S*i is dominant subsidiary and cofunctional. The auxiliary subsidiary *T*e is co-attitudinal with the auxiliary

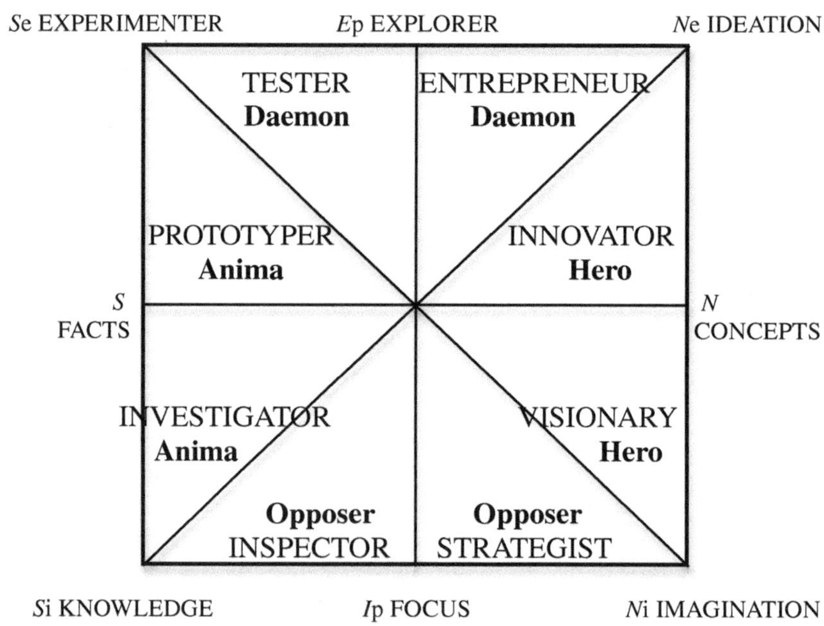

Fig. 8.3 Rotated archetypal perception roles for Example 1

Table 8.5 Male archetype map for Example 5 with *F*e dominant

*S*e	*N*e	–	*T*e	*F*e
Auxiliary **FATHER**	Senex Opposite TRICKSTER		Opposer Opposite *SUBS.* *HERO* DAEMON	**Dominant** **HERO**
*S*i	*N*i	–	*T*i	*F*i
Auxiliary Contravert *SUBS.* *FATHER* SENEX	Auxiliary Opposite PUER (BOY)		Dominant Opposite ANIMA	Dominant Contravert OPPOSER

Table 8.6 Male archetype map for Example 5 with *S*e dominant

*S*e	*N*e	–	*T*e	*F*e
Dominant **HERO**	Opposer Opposite DAEMON		Senex Opposite *SUBS.* *FATHER* TRICKSTER	**Auxiliary** **FATHER**
*S*i	*N*i	–	*T*i	*F*i
Dominant Contravert *SUBS.* *HERO* OPPOSER	Dominant Opposite ANIMA		Auxiliary Opposite PUER (BOY)	Auxiliary Contravert SENEX

mode *F*e. Table 8.6 shows the male gender archetypes for Example 5 with *S*e dominant.

This dominance ambiguity is certainly less satisfying than the clear-cut situation in Example 1, but that's the cost of ambiguity. Sect. 8.5.2 will bring in more information in an attempt to clarify the situation.

8.5 Interpersonal Applications

Archetypes are useful for analyzing interactions between people, giving guidance to team-meisters and managers as well as psychoanalysts. This section presents some examples of this, with evidence that is admittedly anecdotal and experiential rather than scientific. The examples are in fact helpful more for hindsight than for prediction, but they do illustrate how archetypal analysis can give insight into why a relationship is working or not. It may even suggest a correction strategy.

8.5.1 A Two-person Exchange with a Positive Shadow

The *ENTP* author of Example 1 had an interesting e-mail discussion with the Platonically ideal *ISTJ* type professional of Example 4 (Sect. 6.5) that shows how the positive shadow idea can guide understanding of a two-person exchange. It is important in this discussion to realize that the type professional was already quite familiar with Beebe's bimodal archetype theory.

The author asked the type professional his opinion of one of the author's articles. The recipient's reaction was negative at first. He later reported being in touch at the time with his "senex energy", automatically blocking any favorable reaction to what certainly was a deviation from the *status quo*. But he became supportive after an analytical *T*i explanation of the article by the author.

From an archetypal point of view, the mode of discussion introverted thinking *T*i was parentally helpful for the author. For the professional the mode was resolute Subsidiary Parent rather than obstructionist Senex, which would have just deleted the e-mail and forgotten about it. So the professional soon understood the author's point of view and recognized that introverted thinking was for him after all a positive shadow rather than an obstructive Senex archetype.

To summarize: although the professional's first reaction was that of the shadowy Senex, an obstructive old man, his later reaction was that of a "resolute" positive Senex, a subsidiary parent. In this helpful role he gave excellent advice to the author to beware of the widespread taboo on numbers promulgated by most certification programs. This advice certainly influenced the early chapters of the book.

He also remarked that quantitative analysis totally overlooked *F*i and *N*e modes that he experiences strongly. This inevitable consequence of the forced-choice nature of the *MBTI* does, however, anecdotally confirm the archetype model for which *F*i is his *Puer* and *N*e his *Anima*, both in his conscious ego.

8.5.2 A Team Situation with Archetype Rotation

To illustrate archetype rotation, consider the four-person team of which the Example 5 student was a member. Recall that in Sect. 8.4.2 his scores exhibited dominance ambiguity. As a way to resolve this, it is instructive to recount this team's experience. The student was assigned the subsidiary *T*e Organization mode, which he took to with such zeal as to seem rather driven. However, he did not give sufficient attention to the Hero *F*e Community mode to which he was also assigned. As a result his team noticeably lacked unity in comparison with the other teams in the seminar, although his team's somewhat fragmented project performance was on the whole acceptable.

At this point all that can be deduced is that the case of Table 8.5 in which *F*e dominates is the more likely of the two. Unfortunately, the "daring" *T*e appeared to take over from the Heroic *Fe* in this situation. Moreover, the Conciliator role of the *Fe* mode seemed weak, more "speculative" than "masterful' in the words of Tables 8.1 and 8.3. This suggests that the Hero archetype would fit the adjacent *Ej* roles Coordinator *Ej*, *Te* and Diplomat *Ej*, *Fe* better than the Diplomat *Ej*, *Fe* and Conciliator *F*, *Fe* roles of the *Fe* Communitarian mode assigned. In hindsight it seems that it would have been better to assign a teammate to the Conciliator role, relieving the Example 5 student of an apparently opposing responsibility.

Supporting this conjecture is the fact that on a later team he joined, someone else took the *Te* Organization mode assignment, allowing the Example 5 student to concentrate entirely on the *Fe* Communitarian mode. This new team gave easily the best performance of the quarter. Conceivably, archetypal analysis might have warned of a possible problem with the earlier team.

This example raises a new question for teamology that is beyond the scope of this book. Is caution, or at least archetype rotation, needed when a subsidiary mode's decoupled attitude score is so large that its function differs from that of the principal mode? In such circumstances maybe the archetypes in one or both domains are better described by the rectangular axis system, with attitude versus function, than by Jung's diagonal cognitive mode axes. In the example, this would mean characterizing the Example 5 student in the rectangular system as "Extraverted Judgment (Controlling) *Ej*, with Sensing *S*" rather than in the usual diagonal system as "Feeling *Fe*, with Extraverted Sensing *Se* and subsidiary modes Extraverted Thinking *Te*" and Introverted Sensing *Si*". Being based entirely on hindsight in a single case, these speculations must be considered suggestive rather than predictive or authoritative.

8.6 Concluding Summary

In dealing with categorical instead of quantitative matters, this chapter differs noticeably from the rest of the book. It is included mainly to deliver Beebe's archetypal personality model from the oversights of Type Dynamics that were corrected in the chapter preceding. It shows that the archetype model, dealing only with categorical assignments relative to the dominant and auxiliary modes, is perfectly fine when it is built upon the correct dominant and auxiliary modes identified by quantitative theory.

The categorical archetype assignments were conceived by Beebe based both on his practice as a Jungian psychoanalyst and on his examination of his own psyche, much in the way Jung himself worked out the personality theory studied in this book. Consequently the chapter begins by describing Beebe's original formulation, made before the concept of subsidiary modes was discovered.

Once Beebe's archetypes have been sketchily described by keywords, the chapter suggests two ways to expand the theory to include subsidiary modes.

The first, probably more useful for individual personality analysis, is to employ Jung's little known concept of *positive* shadow, occurring when a person represses his own cognitive potential. The second, better perhaps for interpersonal and team situations, involves rotating the coordinate axes of one or both of the mode maps. The latter approach is not numerically quantitative, but it is certainly geometric and graphical.

Although the chapter involves more speculation than certainty, it is included to round out Jung's personality theory for better use by psychoanalysts, *MBTI* counselors and teamologists. It forms a bridge between the quantitative core of the book and the final chapter's review of this quantification of Jung's theory, along with its implications for general personality theory.

8.7 Exercises

8-1 Using your teamology questionnaire results:

a. Construct a table of archetypes similar to Table 8.4 in Sect. 8.4.1.
b. Construct archetype figures, rotated if necessary, similar to Figs. 8.1 and 8.2 in Sect. 8.3.2.
c. As in Sect. 8.2.2, combine mode and archetype keywords to make descriptions for all eight modes.

8-2 If you have *MBTI* scores:

a. Construct a table of archetypes similar to Table 8.4 in Sect. 8.4.1.
b. Construct archetype figures, rotated if necessary, similar to Figs. 8.1 and 8.2 in Sect. 8.3.2.
c. As in Sect. 8.2.2, combine mode and archetype keywords to make descriptions for all eight modes.

8-3 Team-mate *X* of the student in Example 5 has scores 20%*I*, 100%*N*, 20%*F*, 20%*P*.

a. Construct a table of archetypes similar to Table 8.4 in Sect. 8.4.1.
b. Construct archetype figures, rotated if necessary, similar to Figs. 8.1 and 8.2 in Sect. 8.3.2.

8-4 Team-mate *Y* of the student in Example 5 has scores 60%*I*, 100%*S*, 100%*T*, 80%*J*.

a. Construct a table of archetypes similar to Table 8.4 in Sect. 8.4.1.
b. Construct archetype figures, rotated if necessary, similar to Figs. 8.1 and 8.2 in Sect. 8.3.2.

8-5 Team-mate *Z* of the student in Example 5 has scores 60%*I*, 20%*S*, 20%*F*, 60%*P*.

a. Construct a table of archetypes similar to Table 8.4 in Sect. 8.4.1.
b. Construct archetype figures, rotated if necessary, similar to Figs. 8.1 and 8.2 in Sect. 8.3.2.

8-6 On role graphs similar to Figs. 8.1 and 8.2 of Sect. 8.3.2, plot the score points for Example 5 and his three team-mates *X*, *Y*, and *Z*. How would you assign the roles?

Chapter 9
Application to Teamology

Adding a few people who know less, but have diverse skills, actually improves the group's performance.
–James Surowieki, *The Wisdom of Crowds*, 2004

9.1 Introduction

This chapter gives an example applying the quantitative theory to the on-the-spot construction of problem-solving teams from a personnel pool, in this case fourteen sophomores in the author's 2009 Stanford Seminar on "Teamology". Their majors were not recorded, but about half seemed to have an inclination toward science or engineering.

The author, henceforth known as the "Master of Ceremonies" *MC*, orchestrated the activities, supplying the students with questionnaires and mode/role maps as needed. As preparation he determined that every team should have at least four members, which immediately set the number of teams at three – one quartet and two quintets. Students were allowed to designate in advance anyone they wanted to be on the same team with because of friendship, convenience (a room-mate, say), or any other reasons – no questions asked. For the sake of the example, assume that Hanna and Karl declare that they want to be on the same team. Any combination up to the maximum, in this case five. is allowed. No matter what size, such subteams take part in the full team-making procedure, if only to gain information to help them assign members to team roles.

The procedure combines principles outlined in Wilde (2009), updated to take advantage of the new attitude decoupling theory in Chap. 4. It is, however, simpler than the forward-looking methods of *Teamology* because it assigns people in small sequential steps as information becomes available rather than using all information at once. The goal is to have each cognitive mode covered by at least one person who is interested in the mode enough to have a significant score in it. In principle the sequential method used here may be less effective in covering the modes than the simultaneous approach of *Teamology*, but it is also easier to use and understand. In this example, the sequential method appears to do the job well enough. There may be a better mode assignment order; no others have been tried yet.

D. J. Wilde, *Jung's Personality Theory Quantified*,
DOI: 10.1007/978-0-85729-100-4_9, © Springer-Verlag London Limited 2011

The sequential method, whose details are described in the rest of the chapter, begins by having everyone determine their decoupled attitudes from the Jungian *E-I* and Briggsian *P-J* scores they obtain from the five-item questionnaires on pp. 8 and 13. Then they fill out the five *S-N* perception questions on p. 9, using them to determine the students with the top three extraverted intuition *N*e Ideation scores. Each of them then becomes the "seed" for one of the three teams. By coincidence in the example, one of the seeds happens to be in the couple previously declaring themselves a subteam, so they together become the first two members of their team. The Ideation mode is chosen first because its use in an earlier Stanford study, described in Chap. 1 of *Teamology*, doubled the number of teams winning national design awards.

Next the students with the top three scores in the opposite mode introverted sensing *S*i Knowledge mode join the teams they feel most comfortable on. This provides balance on the modal axis *N*e-*S*i, the first to be assigned.

The next axis to be covered is *F*e-*T*i, Community and Analysis, because in the Stanford study this further increased the fraction of award-winners from 1/2 to 3/4. The *T-F* questionnaire needed is also on p. 9. Along the way, people already assigned to teams check to see if they cover the new modes. When they do, their team has no need for a new member in that round.

Finally the remaining axes *S*e-*N*i Experiment-Imagination and *T*e-*Fi* Organization-Evaluation are covered. Any one still left fills any empty openings. Along the way everybody keeps track graphically of their team-mates' scores for future use in assigning team roles.

This procedure takes everyone through the eight modes in a way allowing the *MC* to comment briefly on each mode as it is covered. Time permitting, the roles can also be mentioned. Ideally the students can have some time out of class to reflect on these roles before the teams meets to assign them and begin the first project. In a demonstration during a short course at the 2010 Capstone Conference at the University of Colorado, twenty engineering design professors took a hundred minutes to get through the first five rounds, leaving three quick rounds to place the four people yet to be assigned. All important concepts were covered in this time.

9.2 The Procedure

9.2.1 Decoupling the Attitudes

Once the students have completed the *E-I* and *P-J* questionnaires on pp. 8 and 13, they calculate and record their decoupled attitude scores as in Table 9.1. The results need not be publicized; each person keeps track of their own.

Table 9.1 Decoupled attitudes

Student	Coupled attitudes	%*E*p Exploration	%*I*p Focus	%*E*j Control	%*I*j Evaluation
Aly	20%*I*, 20%*P*	0	0		20
Bea	20%*E*, 20%*P*	20		0	0
Cal	60%*E*, 0%*P*/*J*	30		30	
Dan	100%*E*, 60%*J*	20		80	
Ed	20%*E*, 60%*P*	40		–	20
Fred	60%*I*, 100%*J*		80	20	
Gus	40%*I*, 100%*J*		70	30	
Hanna	100%*E*, 20%*J*	40		60	
Ian	20%*E*, 20%*P*	20		0	0
Jim	0%*E*/*I*, 0%*P*/*J*	0	0	0	0
Karl	20%*I*, 60%*P*	20			40
Larry	20%*I*, 80%*J*		50	30	
Max	60%*E*, 60%*P*	60	–	0	0
Nat	60%*E*, 60%*P*	60	–	0	0

9.2.2 Seed Round 1: Extraverted iNtuition, Ideation

Next use the *S-N* scores from p. 9 to find the top three *N*e scores for the team seeds. Since one of the seeds *H* in this example is part of a declared pair, her partner *K* is also assigned to her team (3). Table 9.2 records the results.

9.2.3 Round 2: Introverted Sensing, Knowledge

The second round uses the other perception function score *S* to identify the top three *S*i scores opposite the *N*e mode. Since student *K*, already on Team 3, is among the top *S*i, the existing dyad *K* and *H* don't need another member in this round. Student *L* immediately qualifies as *S*i because both his *p*-attitude *E*p and *p*-function *S* are positive, while *C* has the highest *S*i score among the three *S*i remaining. Hence *C* and *L* select the teams shown in Table 9.3, and all three teams become dyads.

9.2.4 Round 3: Extraverted Feeling, Community

The next round shifts to the judgment domain, requiring everyone to fill out the *T-F* questionnaire on p. 9. The extraverted Feeling Community *j*-mode is distributed next because in the Stanford study this tactic increased the fraction of

Table 9.2 Seed round 1

Student	%*Ep*	%*N*	%*Ne*	Team
A	0	100	50	1
B				
C				
D				
E	40	50	–	2
F				
G				
H (&*K*)	40	60	–	3
I				
J				
K (&*H*)				3
L				
M	60	*−20*	20	
N	60	*−20*	20	

Table 9.3 Seed opposite round 2

Student	Old team	*Ne*	%*Ip*	%*S*	%*Si*	New team
A	1	*				
E	2	*				
H	3	*				
K	3		<0		20	
B			<0		20	
C			<0		25	2
D			<0		20	
F						
G						
I						
J						
L			50	100	–	1
M		*				
N		*				

award-winners to three-quarters, up from the one-half achieved with *Ne* distribution alone. All of the teams need an *Fe*, and the three *D, F* and *G* are assigned as shown in Table 9.4. Each Communitarian member is intended to make his/her set of individualistic team-mates into a unified *team.*

At this point everyone has completed all the questions and can keep track of their scores graphically on role maps as shown in Fig. 9.1 for unassigned student *M*. As members join teams they should plot their team-mates' scores as well to see where new members are needed as the rounds progress.

Table 9.4 Core round 3

Student	Old team	*Ne*	*Si*	%*Ej*	%*F*	New team
A	1	*				
L	1		*			
E	2	*				
C	2		*			
H	3	*				
K	3		*			
B			*			
D				80	20	1
F				20	20	2
G				30	20	3
I						
J						
M		*				
N		*	*	0	60	

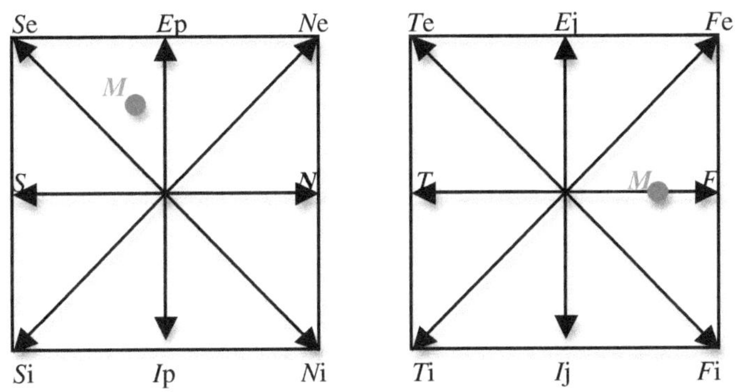

Fig. 9.1 Role map: student ***M*** is marked in *blue*

9.2.5 Round 4: Introverted Thinking, Analysis

For balance, this round assigns the *T*i Analysis mode opposite *F*e. Table 9.5 shows that all teams already have *T*i covered, so no new assignments are made.

9.2.6 Round 5: Extraverted Sensing, Experiment

For no special reason except to continue the assignment process, round 5 assigns the *S*e Experimentally inclined. Table 9.6 shows that the three unassigned *S*e have the same score, so the top one *B* is assigned by a coin flip to fill the only opening, which happens to be on Team 3.

Table 9.5 Core opposite round 4

Student	Old team	*Ne*	*Si*	*Fe*	%*Ti*	New team
A	1	*				
L	1		*		*	
D	1			*		
E	2	*			*	
C	2		*			
F	2			*		
H	3	*				
K	3		*		*	
G	3			*		
B			*			
I						
J						
M		*				
N		*	*			

Table 9.6 *Se* round 5

Student	Old	*Ne*	*Si*	*Fe*	*Ti*	%*Se*	New
A	1	*					
L	1		*		*	*	
D	1			*		*	
E	2	*			*		
C	2		*			*	
F	2			*			
H	3	*					
K	3		*		*		
G	3			*			
B			*			40	3
I							
J							
M		*				40	
N		*	*			40	

9.2.7 Rounds 6 and 7: Ni Imagination and Te Organization

Table 9.7 shows that all teams already have *Ni* and *Te* covered, so no new assignments are made.

9.2.8 Final Round 8: Introverted Feeling, Evaluation

Table 9.8 shows that with the highest *Fi* score, *N* goes to Team 2 and fills the last mode opening. Although they have no gaps to fill, the remaining three *I, J* and *M* each choose the teams noted in parentheses.

Table 9.7 *Ni* and *Te* rounds 6 and 7

Student	Old team	*Ne*	*Si*	*Fe*	*Ti*	*Se*	*Ni*	%*Te*	New team
A	1	*					*		
L	1		*		*	*		*	
D	1			*		*			
E	2	*			*				
C	2		*			*	*	*	
F	2			*			*		
H	3	*						*	
K	3		*		*				
G	3			*			*		
B	3		*			*			
I									
J									
M		*				*			
N		*		*		*			

Table 9.8 *Fi* round 8

Student	Old team	*Ne*	*Si*	*Fe*	*Ti*	*Se*	*Ni*	*Te*	%*Fi*	New team
A	1	*					*		*	
L	1		*		*	*		*		
D	1			*		*				
E	2	*			*					
C	2		*			*	*	*		
F	2			*			*			
H	3	*						*		
K	3		*		*				*	
G	3			*			*			
B	3		*			*				
I									10	(3)
J										(2)
M		*				*			10	(1)
N		*	*			*			30	2

The final roster in Table 9.9 shows that the original mission—to cover every mode on every team—has been accomplished. This is remarkable because this did not quite happen in the seminar, which employed the more complicated approach described in Chap. 4 of *Teamology*. The last three assigned, *I, J* and *M,* are marked with # and considered "wild cards" who, having no formal assignment, are free to fill in where needed. In fact, *M* has two modes *N*e and *S*e where he can reinforce or even replace the others on Team 1 to even out the responsibility.

It behooves each team to use their mode maps to distribute their responsibilities more subtly to cover any gaps in the role pattern, which of course was not taken into account during team construction. "Wild cards" come in handy here.

Table 9.9 Final roster

Student	Team	*Ne*	*Si*	*Fe*	*Ti*	*Se*	*Ni*	*Te*	*Fi*
A	1	*					*		*
L	1		*		*	*		*	
D	1			*		*			
M#	1	*				*			
–	–	–	–	–	–	–	–	–	–
E	2	*			*				
C	2		*			*	*	*	
F	2			*			*		
N	2	*	*			*			*
J#	2								
–	–	–	–	–	–	–	–	–	–
H	3	*						*	
K	3		*		*				*
G	3			*			*		
B	3		*			*			
I#	3								

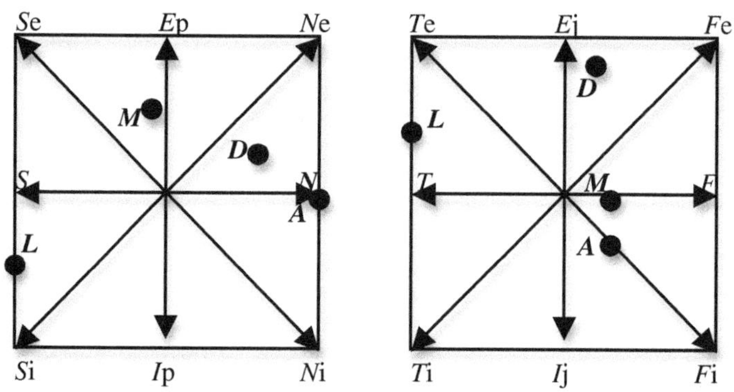

Fig. 9.2 Team 1: role map

Figure 9.2 is an example for Team 1; it has six roles uncovered out of the sixteen total. This is the team discussed in Sects. 6.6.1, 6.6.2, as well as 8.4.2.

9.3 Concluding Summary

This short chapter returns to and updates *Teamology*, where the quantitative theory of this book originated. It provides not only a very practical application to the present book's abstractions, but also a highly motivational way to introduce the quantitative theory to project-minded engineering designers unfamiliar with the psychological genius of Carl Gustav Jung. It also offers conventional *MBTI* personologists entrée into the potentially productive field of corporate teamology as an extension of these academic studies. When tried out on a volunteer group of twenty design engineering professors at the 2010 Capstone Conference, all important concepts and five rounds were covered in a hundred minutes.

Now for the last chapter's overall summary and potential extensions of the theory.

9.4 Exercises

9-1 Add yourself to the example class and carry out the team construction procedure. If you have already taken the *Teamology* questionnaire or the *MBTI*, introduce your scores in the order specified.
9-2 Construct the role map for team 1 as in Fig. 9.2 and suggest a role assignment.
9-3 Construct the role map for team 2 as in Fig. 9.2 and suggest a role assignment.
9-4 Construct the role map for team 3 as in Fig. 9.2 and suggest a role assignment.
9-5 Carry out the team construction procedure for four teams instead of three.

Chapter 10
Extensions and Implications

The most effective way to ensure the value of the future is to confront the present courageously and constructively.
—Rollo May

10.1 Introduction

This final chapter first summarizes the many extensions of Jung's personality theory developed in the book: attitude decoupling, dimensionality reduction, graphical treatment, combination of attitude with function, Platonic quantification, subsidiary modes, unpacking, ambiversion and doubleversion, plus the rehabilitation and extension of categorical archetype theory. Then it develops their implications for other personality theories, namely the *SLIP*, *MBTI* Step II, and Five Factor Analysis instruments. Finally it discusses the disadvantages of forced choice questions with their differential approach to determining personality characteristics. There is a clear need for an integral approach recognizing that people can and do transcend the apparent oppositions built into Jung's personality theory. Indeed, how far can any questionnaire determine presence or absence of the various cognitive modes to the satisfaction of counselors, teams, and individual clients? Many questions are answered, but still more are raised for future research.

10.2 Extension of Jung's Personality Theory

After three chapters of introduction, qualitative theory review, and quantification preliminaries, this book has developed many new quantitative extensions of Jung's personality theory, until now almost totally qualitative despite the development of several questionnaires based on it. Chapter 4 began by using Platonic quantification to decouple Jung's interacting psychological attitudes. This reduced the original single four-dimensional system down to a pair of manageable two-dimensional ones easily represented on simple graphs. This rigorous decomposition immediately exposed the phenomenon of "doubleversion" overlooked by the traditional Type Dynamics categorical approach to understanding personality with the *MBTI* instrument.

Then Chap. 5 showed how to combine decoupled attitudes with functions to generate useful cognitive mode scores. These scores were shown to "unpack" the

D. J. Wilde, *Jung's Personality Theory Quantified*,
DOI: 10.1007/978-0-85729-100-4_10, © Springer-Verlag London Limited 2011

original variables in an elegant way. Brief mode descriptions from *Teamology* (Wilde 2009) were given, abbreviating the richer ones by Haas and Hunziker(2006) offered to practitioners of 8-function(-attitude) personology.

Chapter 6 then uncovered two new subsidiary modes, one in each domain, thereby doubling the descriptive power of the *MBTI*, previously confined theoretically to only a dominant and an auxiliary mode. All these developments were totally theoretical, logical, and mathematically rigorous extensions of Jung's qualitative theory.

Following the quantitative developments, Chap. 7 concentrated on troubleshooting the recently challenged Type Dynamics methodology and providing fixes for it in the two places where they are needed. During this repair process, Loomis' concept of ambiversion was added to that of doubleversion as a way of characterizing the situations in which conventional Type Dynamics and the Myers-Briggs Type Table, although sometimes overlooking creative potential, are at least free from error. Chapter 8 then rehabilitated and extended the earlier categorical archetype theory that had been questioned because of its dependence on uncorrected Type Dynamics. Finally a practical educational application of the quantified theory to team formation and organization was detailed in Chap. 9.

10.3 Implications for Other Personality Theories

10.3.1 Ordinary MBTI Analysis

The quantification of Jung's theory brings several improvements to *MBTI* analysis. The first is to replace the present coupled attitudes with the new decoupled attitudes. This leads to useful two-dimensional graphical presentations and deeper analysis. A consequence is to allow for, and be on the lookout for, doubleversion with its personality descriptions not found in the *MBTI* type table. Another important advance is the detection of subsidiary modes with their attendant creative potentials.

10.3.2 SLIP: Singer Loomis Inventory of Personality

Like the type professional discussed in Sect. 8.5.1, Jungian analysts June Singer and Mary Loomis questioned the forced-choice nature of the Grey-Wheelwrights (*GW*) and *MBTI* instruments, developing their Singer-Loomis Inventory of Personality (*SLIP*) as an alternative. *SLIP* assesses each cognitive mode independently. For example (Loomis 1991, p. 45) the *GW* item "At a party I like to: a. talk; b. listen" is replaced by two items separated in the instrument: "At a party I like to talk." "At a party I like to listen." The responses can range through: "never", "sometimes", "half the time", "usually" or "always".

Conceivably one could apply these ideas to the *MBTI*, adjusting the scoring system appropriately and using the quantitative theory to transform the *MBTI* variables into cognitive modes. The results would be quicker and easier to use than the *SLIP*, although somewhat more laborious than the existing *MBTI*. The advantage would be liberation from the artificial bipolarity of the forced-choice instrument. One can imagine then that the versatile type professional of Sect. 8.5.1 would have, in addition to a positive score for *S*i, a smaller but still significant score for the opposite mode *N*e in the *anima* archetypal position. Thus would be combined the accuracy of direct mode measurement with the convenience of indirect measurement.

10.3.3 MBTI Step II

Some of the same ideas of Sect. 10.3.1 can be applied to the *MBTI* Step II, an important extension due to Quenk et al. (2001). They apply the results of extensive factor analysis of the *MBTI*. Factor analysis involves statistical correlation of the questionnaire items to reduce their extensive redundancy. Factor analysis identifies five "facets" per variable, a facet being a group of similar items whose responses correlate. The twenty items of the questions on pp. 8, 9 and 13 are in fact generated by the twenty Step II facets.

The twenty facets are:

E-I: Initiating-Receiving, Expressive-Contained, Gregarious-Intimate, Active-Reflective, Enthusiastic-Quiet

J-P: Systematic-Casual, Planful-Open-ended, Early Starting-Pressure-Prompted, Scheduled-Spontaneous, Methodical-Emergent

S-N: Concrete-Abstract, Realistic-Imaginative, Practical-Conceptual, Experiential-Theoretical, Traditional-Original

T-F: Logical-Empathetic, Reasonable-Compassionate, Questioning-Accommodating, Critical-Accepting, Tough-Tender

The quantitative theory suggests decoupling the Step II attitudes and refaceting them. The attitude unpacking process of Sect. 5.8.4 would be of help here. There would also be a need for new faceted descriptions of the decoupled attitudes.

Since the items in the questionnaires of pp. 8, 9 and 13 have been arranged in the same order as the Step II facets, it may be possible to identify a subject's pertinent facets. For instance, using coupled attitudes *I* and *J* for a faceted *N*i description would give the following facets for responses to items *EI*5, *PJ*1 and *SN*3: *N*i Imagination: Reflective (*I*), Systematic (*J*), Imaginative (*N*).

From them and pp. 5–8 of the Step II Manual could be constructed the following description of the reflective, systematic imaginative subsidiary mode *N*i for this particular person:

> Learns best in quiet settings. Prefers orderliness. Values possibilities over tangibles.

Things would no doubt be much clearer if a facet description for the decoupled attitude *I*p Introverted Perception were available. Untangling the coupled attitudes would be a formidable task, however, much easier said than done.

10.3.4 Five Factor Analysis (Big Five)

Five Factor Analysis (*FFA*), also known as the "Big Five", is a different personality description system intended more for fundamental research rather than practical application. Largely based on empirical factor analysis rather than theory, its principal proponents are Costa and McCrae (1996), with Block (1995) providing constructive criticism. *FFA*, probably more detailed than needed for teaching or team construction, can be regarded as covering the same psychological territory as the four Jungian mode pairs plus the pair Neuroticism–Emotional Stability, described by Eysenck (1970) as involving depression, impulsiveness, anxiety, hostility, self-consciousness, and vulnerability.

Following are the other four *FFA* variables with their descriptors:

E: Extraversion-Surgency: Warmth, gregariousness, assertiveness-excitement-seeking, positive emotions

O: Openness-Closeness (to Experience): fantasy, aesthetics, feelings, actions, ideas, values

A: Agreeableness-antagonism: Trust, straightforwardness, altruism, compliance, modesty, tender-mindedness

C: Conscientiousness-undirectedness: Competence, order, dutifulness, achievement striving, self-discipline, deliberation

Three others sometimes proposed are introspectiveness, narcissism and forcefulness of behavior. A comparable five dimensions could be Jung's four mode pairs plus the Conscientiousness just described.

One advantage of combining a Jung extension with a fifth separate domain is that the Jung variables are graphable, whereas there is no natural way of graphing *FFA*. An *FFA* difficulty for team-makers, teachers and managers using questionnaires is that self-reporting of Neuroticism and Conscientiousness certainly risks self-serving prevarication. *FFA* advocates seem not to consider Jung's qualitative personality theory and the *MBTI* comprehensive enough for their research. This point of view may change now that a quantitative theory is available.

10.4 A Direct Mode Instrument

Prominent 8-function (-attitude) educators and practitioners Margaret and Gary Hartzler have developed and tested the fsda: Function Skills Development Assessment personality instrument to evaluate the cognitive modes directly without forced-choice questioning. It asks the subject how well s/he uses the skills associated with each of the eight cognitive modes, producing a score ranging from 0 to 6, the top score representing appropriate use with so much reliance on it that it takes little thought.

Fsda scores for the subject of example 1 were: 3*S*e (often), 1*S*i (rarely), 5*N*e (appropriately), 3*N*i, 3*T*e, 5*T*i, 2*F*e (sometimes), 2*F*i. Comparing them with the teamology questionnaire's 70% *N*e. 30% *N*i, 20% *T*i shows that the *fsda* seems to pick up two modes *S*e and *T*e overlooked by the questionnaire if "often" qualifies as equivalent to subsidiary status. On the other hand, since subsidiary mode 3*N*i has the same score as 3*S*e and 3*T*e, maybe none of the three should be regarded as subsidiary. This interpretation puzzle may be worthy of further research.

At any rate, the versatile type professional of Sect. 8.5.1 would no doubt also register significant *fsda* scores for the *N*e and *F*i modes missed by the forced-choice questionnaire for being opposite to his dominant *S*i and auxiliary *T*e modes. Still another way of rationalizing the discrepancy would be to call on archetype theory, which would regard the opposite modes to be third-level *puer* for *F*i opposite to second-level Father *T*e, and fourth-level *anima* for *N*e opposite to first-level Hero *S*i. The different theories here seem to vary only in the broadness or narrowness of their views of the subject's personality.

10.5 Individuation and Non-differential Scoring

In 1991 Jungian analyst Mary Loomis wrote (p. 13):

> In the psychology of Carl Jung, the movement toward wholeness, toward increased consciousness, is called the process of individuation. It...proceeds through the resolution of conflicts, conflicts that are caused by a one-sidedness in our perceptions, attitudes, beliefs, and behaviors.

As people mature, especially when experiencing Jungian counseling or therapy, individuation may cause them to understand and subsequently embrace the modes opposite to the ones they preferred initially. This in turn can reduce the scores for such mode pairs, confusing the use of the *MBTI* as a guide to team construction in the manner of Chap. 9. Thus some of the final round "wild cards" (Sect. 9.2.8) assigned to teams with all their modes already covered may simply be well individuated.

After seeing the teamology questionnaire applied to his University of Michigan Design Science class, mathematical psychologist Richard Gonzales suggested that working with the separate scores—instead of their differences—for each opposing pair variable might identify such individuation. He reasoned that since the teamology questionnaire allows people to choose both options—or neither—separate scores might distinguish between people comfortable with both extremes, who would have high scores, and others undifferentiated or confused, whose scores would be low even when the usual difference scores would be the same.

Thus 100% *E* and 80% *I* would represent someone who has largely transcended the difference between extraversion and introversion, whereas 20% *E* and 0% *I* would be scores for someone undifferentiated in *E-I*. Gonzales' idea deserves mathematical development, which shouldn't be difficult.

10.6 Final Summary

This final chapter first summarizes the book's quantification of Jung's personality theory intended to reveal hidden and unappreciated power of the *MBTI*. Thus decoupling the attitudes reduces the dimensionality from four down to two in each of the two domains, perception and judgment. This allows graphical display in each, which is useful for visualization and team organization. The subsequent exploitation of doubleversion generates more effective teams. Moreover, the detection of the categorical errors of Type Dynamics permits them to be corrected.

The Platonic analysis employed can improve other personality descriptions, namely, the factor analysis based *MBTI* step II and the Singer-Loomis Inventory of Personality (*SLIP*). Further work on the direct estimation of modes is encouraged in the vein of the Hartzlers' Function Skills Development Assessment personality instrument (*fsda*). All of this is in the service of improved detection of creative potential through the quantification of Jung's personality theory

10.7 Exercises

10-1 Use your responses to the questions on pp. 8, 9 and 13 to construct a description of one of your subsidiary modes (Sect. 10.3.3).

10-2 In the style of Hartzlers' *fdsa*, see if you favor any modes not detected by quantitative analysis.

References

Arnau RC, Green BA, Rosen DH, Gleaves DH, Melancon JG (2003) Are Jungian preferences really categorical? An empirical investigation using taxonometric analysis. Personal Individ Diff 34:233–251

Beebe J (1984) Psychological types in transference, countertransference, and the therapeutic interaction, Chiron 147–161

Beebe J (2007) Type and archetype. Part one: the spine and its shadow, Aust Psychol Type Rev 9/2:11–20

Beebe J (2007) Type and archetype. Part two, the arms and their shadow, typeface, 18/3:22–27

Belbin RM (1981) Management teams: Why they succeed or fail. Butterworth-Heineman, Oxford

Belbin RM (1993) Team roles at work. Butterworth-Heineman, Oxford

Block J (1995) A contrarian view of the five-factor approach to personality description. Psychol Bull 117(2):187–215

Brownsword AW (1987) It takes all types! HRM Press, San Anselmo

Cohen J (1983) The cost of dichotimization. Appl Psychol Meas 7:53–249

Costa PT, McCrae RR (1996) The revised NEO personality inventory (NEO-PI-R). In: Briggs SR, Cheek J (eds) Personality Measures, vol 1. JAI Press, Greenwich

Eysenck HJ (1970) The structure of personality, 3rd edn. London, Methuen

Gough HG (1981) Studies of the Myers-Briggs type Indicator in a Personality Assessment Research Institute, paper presented at the Fourth national conference on the Myers-Briggs Type Indicator, Stanford University

Grant H, Thompson M, Clarke TE (1983) From Image to likeness. Paulist Press, New York

Haas L, Hunziker M (2006) Building blocks of personality type. Unite Business Press, Huntington Beach

Haas L, McAlpine RW, Hartzler MT (2001) Journey of understanding: *MBTI* interpretation using the eight Jungian functions. Consulting Psychologists Press, Palo Alto

Harris AS (1996) Living with paradox: an introduction to Jungian psychology. Brooks/Cole, San Francisco

Jacobi JJ (1942, 1973) The psychology of C. G. Jung. Yale University Press, New Haven

Jung CG (1921, 1971) Psychological types. Princeton University Press, Princeton

Keirsey D, Bates M (1978) Please understand me: character and temperament types. Prometheus Nemesis Books, Del Mar

Loomis ME (1991) Dancing the wheel of psychological types. Chiron Publications, Wilmette

Martin CR (1992) Looking at type: the fundamentals. Center for Applications of Psychological Type, Gainesville

McCrae RR, Costa PT (1989) Reinterpreting the Myers-Briggs Type Indicator from the perspective of the five factor model of personality. J Personal 57(1):17–40

Meier CA (1995) Personality the individuation process in light of C. G. Jung's typology. Einsiedeln, Switzerland (originals published in 1977)

Myers IB (1975) The Myers-Briggs Type Indicator (Form G) Consulting Psychologists Press, Palo Alto

Myers IB, McCaulley MH (1989) MBTI manual: a guide to the development and use of the Myers-Briggs Type Indicator, 2rd edn. Consulting Psychologists Press, Palo Alto

Myers IB, McCaulley MH, Quenk NL, Hammer AL (1998) MBTI manual: a guide to the development and use of the Myers-Briggs Type Indicator, 3rd edn. Consulting Psychologists Press, Palo Alto

Myers IB, Myers PB (1980, 1995 edn) Gifts differing: understanding personality Type. Davies-Black, Palo Alto

Popper KR (1934, 1959) The logic of scientific discovery. Basic Books, New York

Quenk NL, Hammer AL, Majors M (2001) MBTI step II manual,. CCP, Mountain View

Quenk NL, Hammer AL, Majors M (2008) MBTI step III manual,. CCP, Mountain View

Reynierse JH (2009) The case against type dynamics. J Psychol Type 69(1)

Reynierse JH (2010) Toward an empirically sound and radically revised type theory (under review)

Reynierse JH, Harker JR (2005) Type versus trait: taxons, real classes, and carving nature at its joints. J Psychol Type 64(6b):77–87

Reynierse JH, Harker JR (2008) Preference multidimensionality and the fallacy of type dynamics: part 1 (studies 1–3). J Psychol Type 68(11):90–112

Reynierse JH, Harker JR (2008) Preference multidimensionality and the fallacy of type dynamics: part 2 (studies 4–6). J Psychol Type 68(11):113–138

Ryle G (1949) The concept of mind. Barnes and Noble, New York

Singer J (1973) Boundaries of the soul: the practice of Jung's psychology. Doubleday Anchor Books, New York

Singer J, Loomis M (1984) Interpretive guide for the Singer–Loomis inventory of personality. Consulting Psychologists Press, Palo Alto

Thompson HL (1996) Jung's function-attitudes explained. Wormhole Publishing, Watkinsville

Wheelwright JB, Wheelwright JH, Buehler JA (1964) Jungian type survey: the Grey–Wheelwright test, 16th edn. Society of Jungian Analysts of Northern California, San Francisco

Wilde DJ (2009) Teamology: the construction and organization of effective teams. Springer, London

Wilde DJ (2010) Psychological teamology, emotional engineering, and the Myers Briggs Type Indicator. In: Fukuda S (ed) Emotional engineering, Chapter 20. Springer, London

Index

FSC
www.fsc.org
MIX
Papier | Fördert gute Waldnutzung
FSC® C083411

Kolibri 360 GmbH
Ferdinand-Jühlke-Straße 7
99095 Erfurt, Deutschland
produktsicherheit@kolibri360.de